From Dragons To Butterflies

Trauma Resolution & Morphic Field Energy Healing

by

Susan Solivan

From Dragons To Butterflies –
Trauma Resolution & Morphic Field Energy Healing

Cover Artist

My Thanks go to Susanne Shapp-Valla for her beautiful rendition of the art piece I had requested some years ago. We had traded sessions for her artwork, and I had simply asked for a drawing of a Dragon, a Girl and a Butterfly. She couldn't have done a better job. The Dragon is a water dragon, and my book is about emotions. The girl is held and protected by the dragon and in such a sweet space that a butterfly is about to land on her hand. The pastel colors are exquisite, peaceful and light. It has been many years that I have enjoyed her artwork in my home and now I am proud to feature it on the cover of my book.

Susanne can be reached at slvalla@hotmail.com

Thank You, Susanne.

In Loving Memory Of Susan Solivan
1946—2014

"Though you're far away, I have only to close my eyes and you are back to stay, I just close my eyes and the sadness that missing you brings soon is gone and this heart of mine sings."

Ernest Solivan

Table of Contents

How to Use This Book

I began a monthly series of talks at a local gem store, Chrysalis, owned by Kathleen Rodamere here in Merritt Island, Florida where my husband and I now live. I was teaching Morphic Field Energy Healing and Trauma Resolution.

After some time, the participants began to successfully work with morphic fields and had no trouble 'feeling' the twisting and turning and finally the balanced resolution of the morphic field they were holding. There were such positive results from the work, and the participants were faithful to the class attending when they could to experience more practice. I demonstrated a few Trauma Resolution sessions but for teaching that we needed more time and sacred space of a different level.

The idea to have longer classes with more people at additional locations began to take form. This book is written to enhance those classes. The chapter titles and information contained within is just an introduction of what is important to understand before you take on any of the work addressed here.

I have been working on myself and clients since my late 20's and this year, 2013, I turn 67. I finally understood that all the work I have done with clients was really work I did on myself. Many of my clients have been my best teachers. It is important for you to understand that every interaction you have in your life is your creation and there is something to gain in it. SEEING it with clear vision is most important. For me having clients to reflect or mirror my own issues was a real advantage and addition to my own healing.

Lynne Forrest's *Beyond Victim Consciousness* and Jack Carson's *Gremlin* series is most important to study, read, and act upon. Watching *What the Bleep* is a film that you need to watch over and over to get that we are all One and we create our own reality.

Jane Robert's channeled work, one of which is *The Nature of Personal Reality* will open your mind, I promise you. It is also important to understand the newly discovered Parallel Universe (See Brian Greene) in quantum mechanics to understand that everything is happening now AND that any road in our life we did not take is continuing in another parallel universe. (Jane Roberts, 1970) Dr. John Bradshaw's work on Family Dynamics will open your eyes as to how your birth family impacted your habits, thoughts, and behavior patterns today.(See www.johnbradshaw.com)

There is no need to be overwhelmed. Use your gut feeling about where to begin your study of you and your life. The more you are clear yourself, the more you can help others. The most important is of course, that you take care of yourself first, before any client, friend, or family member. Fill your cup before you can be present for others, otherwise you are empty and cannot be truly present for yourself and your life, let alone your clients. The drain you will feel from not filling your own 'cup' will most likely drive you away from your work or else you will end up being a Star Victim. (See Lynn Forrest)

In other words, this book is not the end all be all for getting your head on straight or at least becoming aware that it's a bit off kilter. This is a short book touching on this and that as invitations to go deeper so that at some point you can BE in the world as the full, luminous, and bright soul that you are. Sometimes my lights dim or go out a bit and I need to reconnoiter and get grounded again to be aware that I am connected to everything and everything is always in Divine Order even though it may not look like it.

Therefore, if you find the subject matter and practices of this book interesting, it is but a taste of what is out there for you. I encourage your journey, and my wish is that it is full of everything that is best for you. Remember, we all create our own reality. It is up to you to bring in the best.

Blessings to all,
Susan Solivan

ADVISORY TO MY READERS

In so much as this book is about healing Morphic Fields and Resolving Trauma, it is ill advised to begin to set up a practice with just the small amount of information contained in this book. My personal stories and client treatments would not be enough for anyone to jump start a consulting business with only the information contained here.

The reason I included so much information on dysfunction and personal and interpersonal issues is to advise the reader who might hope to do this work that doing their own personal work and understanding dysfunction in their family patterns will not only allow you to remedy and clear those patterns but be more aware and present for your client's similar issues.

If you jump into this without seeing your own 'stuff' it will come up in possibly very uncomfortable ways. At least begin your own personal work to See Clearly how your issues are mirrored by your client. Of course, it is advisable that you get complete training either in BioDynamic CST, (BioDynamic Training, also referred to Breath of Life), Somatic Experiencing ©, Bodynamics.com or consult SpiritRoc.org for extended training. If you are already a trained professional, the

Healing Trauma method explained and referenced in this book is quite different from other approaches to psychotherapy.

I recommend researching, reading, and investing in clinical training that would assist you in becoming the best you can be for anyone coming to you for treatment. This book is no substitute for medical or psychotherapeutic treatments.

I have included a list of what I consider to be pertinent and important books that can only benefit your own life and further education in dealing with the issues and subject matter contained in this book.

I wish you the reader every advantage in reaching any goal you may have about the material contained herein.

Respectfully yours,
Susan Solivan

ACKNOWLEDGEMENTS

First and foremost, I wish to thank my birth family because without the dysfunction I experienced, I would not have gone onto this path of self-discovery. They have all passed at this time in my life and my hope is that if indeed they moved onto another life that it be easier and clearer for them to get to the core of their connection with Universal Mind, Source, God, All That Is and to realize that they too are an integral part of the Whole that Is. We are all inherently perfect and beautiful.

My mother was a book lover. I began reading the volumes she supplied me with as a child and fell in love with books that were able to bring understanding into my tumultuous life. From fairy tales to spiritual books to quantum physics and everywhere in between, I found books to be my teacher and friend. I did not have a classical education so the many wonderful tomes of fun, understanding, and information became my favorite pastime.

You will find a list of some of my favorite and pertinent books mentioned throughout listed at the end of this book. It is truly said that when the student is ready the teacher appears. That has been true for me with books and teachers as well, including any relationships I experienced over the years.

My husband Ernest is just what his name means, earnest in all things. His calm approach and acceptance of whatever I am going through has been my rock, my fulcrum around which the storm of my healing found a peaceful tether. He has never once criticized me, only told me daily that he loves me. I think after almost 30 years, I might finally be "getting" that.

My children were and are also my best teachers. Everyone we meet in life is a mirror, showing us ourselves. It is an amazing gift if you can see it this way.

All my teachers over the years are too numerous to be mentioned here so I will name the man who brought me to myself with all his classes and his quiet acceptance of who I was at any given time. Dr. Michael Shea is a worldwide teacher of BioDynamic Craniosacral Therapy. His teachings confirmed the trauma work I was already doing with clients and better yet, got me to begin to see the shortgivings that were clouding my life.

The BioDynamic approach to CST (Craniosacral Therapy) is one that calms the nervous system and allows integration of trauma wherever it is held in the body in a peaceful, quiet and healing experience. The many hours of training and class time have brought a bright horizon into my life. I was fortunate to have met some amazing therapists through the classes we shared. (See www. michaelsheateaching.com)

There are no coincidences in life and the way I 'stumbled' onto Dr. Shea is nothing short of a miracle. I love this story and will share it here. I was living in Arizona and practicing as a massage therapist as I did in Florida in the 1980's. I had just finished another Visceral Manipulation class and decided to ask a few other practitioners along with Dr. Barry Malina if they would like to be in a study group to confirm what we had learned with mutual practice.

Dr. Malina taught A & P at the osteopathic college in Arizona so was kind enough to head up the anatomy, palpation, and visceral treatments during our study groups. Soon after we began, he said he was excited about a class he was going to attend with another osteopath, Dr. Jim Jealous. When he returned from the class, BioDynamic Craniosacral Therapy, we all could see what an amazing new CST approach to the body this new therapy held.

Some weeks later, I was speaking to another massage therapist friend of mine in Florida and asked her if she knew of anyone who taught Biodynamic CST. She said I should talk to Cathy Shea, a colon therapist and teacher in south Florida. When I talked to Cathy the next day, she said her husband, Dr. Shea, was just beginning teaching this modality. He agreed to come to Arizona if I could get 15 interested students. This was sometime in the mid 1990's. From there, not only was I privileged to do the three-year training of 750 hours but in addition I attended many more BioDynamic CST classes from then till now.

I hope you can see the incredible series of events that occurred for me to get this training that has become so valuable in my own healing. There are no coincidences. It was as easy as calling a friend to have found this man through a few series of connections. Dr. Shea's brother is an osteopath and trained under Dr. Jealous. My thanks also go to Dr. Malina who was kind enough to teach a few classes on the BioDynamic model. He continues to practice in Arizona.

Whatever has happened in my life I truly believe that I had foreknowledge of all of it when I decided to incarnate into my birth family. I could 'see' the lessons I would learn and experiences I would have that would enrich and challenge my life. We do indeed create our own reality and there is no one to blame, only we ourselves to take up the banner of our own individual lives and receive each day as a gift.

Chapter 1

QUANTUM PHYSICS

Yes, that is a daunting title for this chapter. This chapter explains how QP interfaces with the psychic and healing experiences. No, I will not be going into string theory or mathematical formulas, so please do not be intimidated. However, I have dabbled a bit in some pretty good books that are lay person and user friendly.

This was all done in my quest to understand the How of me being able to influence and read a client while on the phone or Skype across the world or down the street. I had to understand how it worked and when I had a glimpse of the dimensional QP involved, the work I was doing became more effective because I began to understand how that these things could happen. Once you understand how something works, it becomes more available in a deeper way.

I think the very best and most comprehensive demonstration of how quantum physics affects our lives is best explained in the movie *What the Bleep!? Down the Rabbit Hole.* It is one you want to watch over and over as it will always renew your interest and understanding about this dimensional world that has so much influence on our existence, whether in the physical or the spiritual.

At first when it was so easy for me to do readings on clients and see in my mind all about them, I remembered my catholic upbringing which condemned psychics. In fact, psychics, called witches back in Salem days, were burned at the stake for being *supposedly* in league with the devil. It was a breath of fresh air for me to let go of that belief and begin to understand that there was a solid math and quantum physics background which explained what I and others could do.

I can take the mystery out of what I do so that clients do not revere me or put me on some idiotic pedestal because I can do something that they cannot do. Contrary to that, everyone can do this work if they can learn to ground themselves, meditate, and have a heart's desire to help others.

As I look back, it felt like some higher power, probably me, was setting up just the right scenarios and teachers for me to do this. It is not for everyone so if you have no interest in developing your abilities so you can help others, no worries. There is another path to follow that is more suited to your energies. However, being able to be in your heart space and to tune into situations and get gut feelings for events will enrich your life and allow you to always feel more balanced.

As the quote reads, "If the student is ready, the teacher will be there" and that has proved true for me. Trust that as you read this book and if your interest goes beyond what I have written, know that the perfect set of situations or the perfect teacher will come your way.

As you read in my *Psychic States* chapter when I first contacted someone who had died, the idea of other dimensions was obvious to me. Since then, Burt Goldman (Google his very informative website) and Brian Greene's last book *The Hidden Reality:* **Parallel** *Universes and the Deep Laws of the Cosmos* both demonstrate unequivocally that there are other dimensions.

I read a very intriguing book sometime in the '80's called *Flatland* written in 1884 by Edwin Abbott. *Flatland* is about two-dimensional creatures that of course cannot see beyond their two dimensions. When one character sees a shadow beyond his dimension, it causes a furor within his group of other two-dimensional beings. This is a small book and well worth the purchase for the mental stimulation and understanding of other dimensions that may be unseen by us as they were in *Flatland*.

As the years went on and I had contacted more and more people who had died it became very clear that they still existed but in another dimension. There is work they do there, learning or even teaching. The physical personality who existed previously and currently is in another dimension (as having passed from their physical life) is now working on other projects relevant to their new dimension and abilities.

Frances

When I lived in Ft. Lauderdale, I had a client who passed away from cancer. About a month after she died, I was driving down Commercial Boulevard with the windows down inviting in the beautiful breezy day. I was in bliss, peaceful and smiling. Suddenly, I 'saw' Frances in my mind's eye and at the same time heard her say that she was so glad she finally passed saying into my mind, "Everything is so beautiful!" Without saying another word, she sent me feelings and sensations of the exquisitely beautiful and breathtaking place she was now experiencing in her new life. This all came into my right brain and my body at one time with a rush of ecstasy that made me gasp with a huge intake of breath and brought me to instant tears of joy. Notice that I was in my heart space, peaceful, happy, and in bliss, enjoying the day. Otherwise, Frances would not have been able to send me that message. I think our minds are always busy thinking, worrying, projecting, and filled with thoughts that do not serve us nor let us relax.

I am easily able to drop down into relaxed brain waves. We are normally on Beta, the faster brain waves of our daily existence. Calming yourself down and slowing brain waves can bring you to Alpha which allows information and pictures to come in. It's kind of like having a busy signal all the time in your brain nothing can come in because it is too filled with 'stuff'. There are people who can go to Theta and Delta brain waves, very deep and serene. We vacillate in and out of different brain waves during our sleep cycles. Suffice it to say that Beta is Busy and Alpha is Available.

Quantum Entanglement is another Quantum Physics mystery to explore. Roughly put, it states that connections that we feel that appear only physical and in that moment are not the only types of connections that can happen. Connections can be non-local, which is why if someone is thinking about you, you can pick that up and think about them at the same time even if they are on another continent. Everything is connected, everywhere.

Every client or person or animal I ever encountered in my life is connected to me. And, even more profound everything is connected to everything else since we are all one and there is no time or space. We only need time and space to live within this physical reality. If you want to hear about protons splitting and knowing its other proton (Quantum Entanglement) then you can Google the real deal.

Example of Non-Local Connection & Quantum Entanglement

Some years ago, I worked in a billing office. I was on my way to work when I was hit, rather blasted, with head, neck, and shoulder pain. I knew it belonged to someone else (because it came on so suddenly) so I planned to sit in the cafeteria watching who would come in for breakfast before office hours. Sure enough, someone in my division came in and bingo, I knew it was her. My body recognized the pain in her and instantly the replay of her pain left my body. I approached her when we had our breaks and asked if she was in pain, yes, she answered. I told her I could probably lessen it if she were willing. Yes again. We sat in a closed office, and I did energy work for about 10 minutes, and her pain was gone.

This was Friday and when I came into work on Monday she wasn't there. My supervisor told me she moved back to Canada and took a job in a doll making factory. Here is the story. She had such a bad auto accident that she received compensation in the millions. Therefore, if she was pain free, how could she justify having all that money. She took a job in a doll making factory which meant she would be using her hands and bending her neck doing the work of putting the dolls together. All her actions were below conscious awareness. In this example you also may note the non-local connection that happened as I 'got' her pain driving to work that morning.

According to Amir D. Aczel's book, *Entanglement, The Great Mystery in Physics,* "Whatever happened to one particle would thus immediately affect the other particle, wherever in the universe it may be. Einstein called this "Spooky action at a distance." This succinct quote may give you a better understanding into the non-local mind experience.

Connection

For me, it is simplified by the following experience of my connection to my clients over the years. A good example happened some years ago with a woman who came to me to understand the relationship with her husband. I would tune into how he saw her and explain so that she could go home and do better with their relationship.

This happened a few times until she then felt her relationship was on better ground. I was later invited to a gathering at her home, and her husband came up to me and kept saying, "You look so familiar, are you sure I have not met you before?" He kept staring at me off and on all night, of course we could never tell him that I knew him through a psychic connection. When his wife and I had gotten together, we brought the essence of his morphic field into our presence and thus on some level, he was there also.

As an interesting aside I think many of you have heard the expression, "Quantum Leap" in regard to someone getting ahead in a big way. What it really means, according to Dave Jarvis (2010) is: "Quantum leap is the act of an electron jumping between 2 shells while orbiting a nucleus." If you can imagine how impossibly small that action is one would wonder why that phrase is attributed to a huge jump in our human lives. Perhaps to the electron it IS a huge leap.

Another writer who has influenced my work is the theories of Rupert Sheldrake and Morphogenetic Fields. (He is mentioned quite often in this book) More recently I attended a series of Matrix Energetics seminars hosted by the developer, Richard Bartlett, D.C., N.D. I watched him onstage run his hand down the front of the volunteer after he asked what the problem in their life was and he lodged his hand in front of the heavy feeling field he found as he went down the field again and I could see he collapsed the field (which is only in-FORM-ation, you MUST please understand!) and the volunteer would fall over.

The explanation is that so much was removed from his field that it would overwhelm the physical body, and it would collapse. Assistants would stand behind the stage volunteers so that as they fell over they were gently lowered to the floor.

When one thinks a thought or a memory, the morphic (or energy) field of that mental moment is surrounding the person involved in that thought. If it has a strong energy field, it can also be perceived by someone who is sensitive to energy.

One might think this is all staged, but I have felt it both as a recipient and participant. It may be localized in different areas of the physical body which is why the therapist will run their hands slowly through the field to find where it is. Richard spoke about morphic fields and how we carry them around and the more you talk about them, the stronger they get into your body and field. Say for instance you fell onto a concrete sidewalk in front of a favorite restaurant.

Each time you remember that fall or are in front of that restaurant, the morphic field of that fall shows up as a memory and can impact you almost as much as the original fall itself. More on this is in the Morphic Field chapter. It is well repeated here to firm up the understanding of how powerful these fields can be.

Once the memories are integrated, for instance, using trauma resolution or morphic field clearing, the trauma loses its charge and becomes only a memory. At that point, remembering the incident does not bring up dissonant energies in the field around the body. (See the chapter on Trauma Resolution and the Instruction chapter to understand how to clear, release, and integrate morphic fields and trauma)

Non-Local Mind and Prayer

You can Google Larry Dossy, M.D. as he speaks eloquently about non-local mind. He encourages prayer groups who work at a certain time in unison, whether they are in Africa, America, or the Netherlands. It only matters that the prayer be all at the same time. His goal is to show how powerful prayer can be. In one experiment there were groups who worked on Washington, D.C. where the crime rate was off the charts. There was a remarkable drop in crimes during and after the prayer groups.

I see prayer as filled with Light. Good thoughts aimed toward another are as good as prayer, actually the same as prayer. It does not matter what deity one prays to, it is all Light and part of Universal Mind or God. My friend Roger had a mom diagnosed with a serious heart condition. She was slated to have a short time left. He got a group of friends and church members of hers to pray for her and the heart problem was no longer in her body! She lived at least 15 more years. I could wow you with so many true stories like this. Do not discount good thoughts toward others or prayers. The Light that is Source or Universal Mind or God loves to have its Light offered to others. By the same token, negative thoughts can be hurtful, especially to the person wallowing in them.

Experience

Some years ago, I knew of a woman who was very angry with me. I received a phone call to say she was coming over to my house. I knew she would always regret doing that so I sat down in my quiet house by myself, and I prayed that she would be somehow stopped. I was calm and sent love her way showing a higher part of her that this action would serve no one.

The next day I went to my friend Melody's house. She told me about a woman who was very angry and was going to do something she regretted to another woman until her neighbor came up to her and stopped her, letting her know it was not a good idea. It turned out that Melody and I had worked on this neighbor woman for years, as she had had a stressful life and relationship. (What were the chances that she was the neighbor of this woman who was angry?) Do you see that my connection to the neighbor (who I had never formally met in person) and the prayerful wishes I sent that way to stop the angry woman were carried out by the neighbor who knew me only as a non-local connection. I am still amazed by that even though I am quite aware we are all connected and can influence others with our good thoughts.

I remember the catholic teaching said that gossiping was a venial sin. I later saw that gossiping can be energetically hurtful to the one being gossiped about. Like the old saying goes, "If you don't have anything nice to say, don't say anything at all." Usually, criticism directed toward another person is something you don't like about yourself or are afraid of acting out in the same way.

Chapter 2

THE SUBCONSCIOUS MIND

Sometime in the mid 1980's, I signed up for a hypnosis class. I had just finished massage school and discovered that if the therapist says something negative during the session, it can go into the subconscious mind and activate in the same negative manner in the client. As an example, say the person has a sore shoulder and the therapist says something like 'oh, this is the worst shoulder I have ever worked on, I bet you have had trouble with this for years! This may never get fully healed.'

Because the client is a bit dissociated lying vulnerable on the massage table and suggestible because certainly this trained therapist would know more than the client, or so they think, the idea of a 'really bad shoulder problem' can become seated in the subconscious mind of the client who then will have continuing problems with that same shoulder.

Conversely, the therapist can find something positive to say that will add health to the client, as example, "I can see this has been hurting for a while and I am glad you are here today for shoulder work. Between the restorative health of your body and this massage work, your shoulder should get progressively better as we work together."

It certainly seemed important that I examine this subconscious 'suggestion stuff' so that I did not do more harm than help for my clients. This small interruption in my learning curve turned into one of the best investigations to help myself and others. It turned out that this seemingly innocuous little part of us, our subconscious mind, is an intensely powerful and deciding factor in how we live our lives.

Let us start with childhood. To the children, their parents are all powerful, first, and then of course, all knowing. The child must assume this as s/he knows these are the people s/he depends on for care. Before the age of 7, a child has no critical factor (every word the child hears becomes truth and goes into their mind without being filtered) and cannot discern what is truth or untruth. So, if the parent(s) tell the child s/he is stupid, worthless, a huge bother, etc., the child will take that in as truth. The child may not remember the words, but the feeling of that put-down will be forever in their psyches. (See Judith Herman, *Trauma and Recovery* It's about inappropriate and damaging power over children) Even so called teasing in 'fun' can be damaging as the child even takes teasing as truth.

Further, even in adult relationships such condemning and critical name calling can last a lifetime in the subconscious mind of the child or adult. As life goes on they might wonder why nothing ever goes right, why their jobs are just menial, why they can never seem to get ahead and so on and that is because their programming of being called 'stupid' and worse is running in their subconscious computer.

How can this be remedied? There are several different avenues for help. It is important that one be sincere about working on such issues and look for a modality that fits. I mentioned hypnosis as a good one. It is important to find someone experienced, even a trained hypnotherapist is a good choice. My husband works with the left and right hemispheres of the brain in his Hemispheric Kinesiology sessions, and I use Trauma Resolution work. There is also EMDR (Eye Movement Desensitization and Reprocessing), EFT (Emotional Freedom Technique), BioDynamic Craniosacral Therapy, Jungian and other Psychotherapies, to name just a few.

The subconscious mind is a powerful tool and participant in our lives, for good or ill, depending on the circumstance(s). No one pays much attention to it or if they do, it is not considered an important subject to discuss. Here are some things I have discovered about this part of our mind. From here on in, we will name it the SC mind.

The SC mind holds all memory of what has happened to us and will protect us at all costs. Here is a for instance: let us say that as a child you had to remain quiet, invisible, and submissive to be safe in your abusive household. The SC mind protected you with those habits and unless you change them consciously, you will be a quiet, invisible, and submissive adult. If you notice that it is a problem and decide to clear any issues relating to that, I assure you that your life will improve, even if you have a rocky road for a while during the resolution of the memories.

If your feelings as a child were that you were worthless and didn't deserve to breathe the air around you or hold space in the world, you might end up with an eating disorder and never understand why you have a compulsion to starve yourself into disappearing.

If you have a fear of bridges and you start thinking about a bridge that scares you, your body will react to that very fear as if you are there. The SC mind does not differentiate between real experience happening in that moment and thoughts of some trauma in the past. If you stop to think a moment, you may realize you have been terrorizing yourself for years thinking thoughts of the past frightening experiences or imagining worst case scenario in the future.

By the same token, this is the time to do good work with the SC mind and clear some issues. When we have any kind of trauma whether when we are a child or adult, the SC registers the experience using the body as a storage place.

Dissociation

We all dissociate throughout any given day. During a task that must be completed quickly, usually the right brain shuts down, at least somewhat. The same with the left brain closing down if you are sitting and looking at a sunset. The detriment comes in when one or the other, and sometimes even both, sides of the brain shut down for long periods of time.

There is also a Dissociative Disorder where both sides shut down, without return to normalcy. The Google/Wikepedia definition reads: **"Dissociative disorders** (DD) are conditions that involve disruptions or breakdowns of memory, awareness, identity or perception. People with dissociative disorder use dissociation, a defense mechanism, pathologically and involuntarily. Dissociative disorders are thought to primarily be caused by psychological trauma."

There are cases where patients lose the ability to speak, hear or see, and possibly even appear in a catatonic state, not moving or responding because of a severe trauma.

I recently heard of a severe trauma to a mother and father where their very young child was brutally killed. The mother is appearing zombie like, barely making it to work and having little interaction with anyone. She is severely dissociated and frozen (Levine) in the unbearable moment she heard about her son. Her deep grieving is unable to begin because of her freeze.

This woman's experience is at the extreme end of shock and trauma. It will take time with a very good therapist to help her come back to her life. This experience was too enormous to take in. It is a huge field surrounding her, waiting to be integrated. She will need an astute therapist to see the right time to do this with her patient.

<u>You cannot get through trauma without going through the body!</u>

The left hemisphere of the brain is about details, remembering the past, rigidity, being critical (of anything), and a bit cold in the feelings department. The right hemisphere is warm, loving, sees the big picture, has no boundaries, and 'anything goes'. So let us use an example of someone addicted to something. It doesn't matter what, an addiction is an addiction.

What happens here is the left brain shuts down and the right brain says everything you do is ok, no regards for anything but what is in the moment to do. The OCD or compulsive person who must be controlling, always on time, critical of everyone including self, has their right brain shut down. Something (trauma) happened at some point in their lives to compromise the balance between left and right sides of the brain.

How does this show up in someone's life? If it is too scary for the person to face a past trauma, depending on how it impacted them, they will switch off one or the other side of the brain. We have even seen both sides shut down…that is a huge out-of-body dissociation where anything goes. Usually, the person has a disconnected feeling and may even become isolated because they cannot connect to anything or anyone. At the moment of dissociation, you can look at the eyes and see the vacancy; no one is (literally!) home!

I think this is how heinous crimes happen; the person is acting out and totally dissociating. No, that doesn't excuse the crime. The dissociating is the thing to cover up what is really going on deep in the psyche. I remember an article in the paper about a man exposing himself to a group of children in a playground, surrounded by teachers and older students. Of course, he would be arrested but his compulsion was so strong he dissociated, did the deed that if he were conscious he would know he would be arrested and of course that was the result. He may later be confused thinking "why did I do that?"

Someone who feels empty inside is going to try and fill that up with something, food, drugs, or even excessive work habits, anything to stave off feeling so empty. They may find themselves looking for love, as they say, in all the wrong places. This can show up as going from one sexual partner to the next, all the while the SC is hoping to find someone who will love them. If the SC blockage is that they are unlovable, they will never find that person to love them. What we all see in acting out behavior is simply an out-picturing of some trauma or hurt that the individual is trying to cover up.

According to some new therapies that are coming to the fore, there is a way to work with the mind and body to bring resolution to aberrant behavior. And of course, the individual would need to want to change. The best therapist/therapy on the planet doesn't work unless the person is willing to work as well.

The Limbic System

I will be mentioning the limbic system off and on and briefly it is a set of brain components which are important in emotional responses, drive-related behavior, and memory. (*The Human Brain*, Nolte)

The 'components' are: the Amygdala, relating to emotional states, offers quick response to stress and/or danger. (Sills, 2001) The Hippocampus: relates to the memory system and the olfactory system (sense of smell) (Upledger, 1987). The Hypothalamus: controls the autonomic nervous system within the limbic system. The Pituitary gland secrets 9 hormones for homeostasis. The Pineal gland also works with the endocrine system producing melatonin for sleep, sexual libido, and converts nervous system signals to endocrine signals. The pineal is also the visionary area of the brain (see the book on DMT) and when developed provides mental pictures of amazing clarity.

If you remember what lobotomies are, or were, it was a surgery to remove the Amygdala. Once that was done, the patient no longer had any emotion, also called flat affect. It was a way to control patients housed in mental institutions. (see the movie, *One Flew Over The Cuckoo's Nest* with Jack Nicolson) Yes, it seems extreme, and I believe that is now not practiced.

Any kind of abuse or trauma can cause problems in our system and our lives. Some of which are: hypervigilance (always being alert for danger) which stresses the adrenal glands and shows up as constant anxiety; MPD or DID disorder (multiple personality disorder or dissociative identity disorder), where different personalities take over other personalities developed to protect the psyche of the survivor; addictions and compulsions; memory loss of the trauma(s); PTSD or post-traumatic stress disorder, common where a child or adult was exposed to extreme trauma, usually life threatening. And so, the list goes on.

The SC mind holds all the personalities (in MPD or DID) and allows the one that can help in any given situation to take the lead when the person needs help or feels threatened. It is an exquisite survival system that allowed the traumatized person to live through some terrifying event. It is not a bad thing and at the same time, carried forward into adulthood when the trauma is gone, can cause problems in the life of the adult.

All the while, the body and SC are holding all of this and at the same time running away from it (the old trauma) with mental anomalies and strange behaviors. Because of my own personal history and that of my family members and friends, I have become intimately acquainted with these difficult situations.

As you read through these pages, you will see much repetition and different explanations or methods of assistance. I have listed some books that may be of help and will be explaining some of the methods I have been using. This book is only an eye opener and the beginning step to a road of understanding and healing.

Learning to Be Embodied

Self-regulation practice is one that we all can use. Instead of blowing your top, you can slow down your reactions and feel in your body what is going on and then decide about what to do that will settle the situation rather than add fuel to the fire. If you are easily activated (the smallest thing can unsettle you, make you angry or scared, step up your heartbeat, make your hands sweaty and so on) practicing self-regulation will allow you to face these activations with little difficulty. When you notice your body beginning to activate and can ground yourself quickly, you will be able to stop the activation before it gets out of hand.

I remember a newspaper story about road rage between two women. One woman got out of her car with her gun and shot and killed the other woman. Immediately she was filled with regret and to this day is devastated to have taken the life of another human being. IF she would have been practicing self-regulation she would have calmed herself down and been able to deal with the situation in a more supportive way for both of them.

You are at a party with friends, and someone gets out of hand because they have had too much to drink. They get loud and abusive to someone nearby and you notice your mouth gets dry, your hands begin to sweat and you feel weak. This is keying a body memory for you and instead of going into a full panic attack or being trashed the rest of the week, you can notice that you are feeling very uncomfortable. Before you go to full blown activation and panic mode, take care of yourself.

The best thing is to get away from the situation that is activating you and go to a quiet place where you can ground yourself and notice that you are calming down. Breathe slowly and fully into your body and regulate yourself. If you are seeing a therapist this experience will need to be mentioned in session, but it is up to you in the moment to learn how to self-regulate. The practice for this can be found in *Healing Trauma* by Peter Levine.

Remember the practice I advocate of getting stable and in your body and then mentally going somewhere else, it doesn't matter where, just mentally go out. When you come back notice the body sensations and then leave again repeating the grounding sensations. Once you do this enough times, you will have that strength available whenever you might need to get stable and grounded. It is like a pendulation, going back and forth from dissociation to being grounded. The result over time will allow you to quickly notice when you are 'out' of your body.

Frozen Shoulder Thaws Out

A massage therapist friend of mine sent me a client with a frozen left shoulder and arm, 6 years running. Rosita had tried deep work, but it was too excruciating. One of my modalities, as an offshoot of craniosacral therapy, is NeuroFascial Release. This involves the therapist laying their hand or fingers on the point of pain and then sensing the other end of the lesion, or pain link. Within a few minutes of sitting in this position with my client, she began to yell that it was so painful. Now, physiologically, the fluid in the tissues had become viscous, thickened from lack of movement. I could "see" it as a sort of dried egg white and while I was holding these two points, the fascia was responding by getting the fluids to move. I let her know we were doing that, and the pain would resolve soon.

A few minutes into this, I heard in my head, "empty arms". She had told me she lost her son when she divorced her husband and had fought the courts for years to see him, but her lawyer ex-husband knew how to block her at every hearing. I asked her if I could say what I heard in my mind and that it may cause a reaction. She agreed and I simply said, "Empty arms".

Her body immediately started shivering and she got very cold. She asked what was going on and I explained that her body was releasing some sort of trauma and she would be fine in a few minutes. I wrapped her in blankets as she was still on the massage table and waited until it was over. She then also told me she'd had an abortion when she was very young and felt as if she had lost yet another child with her second one.

As she calmed down, I helped her off the table and when she got off the massage table, her frozen arm and shoulder were totally free. Before that she could only lay her forearm on her stomach and barely move it to shower or do tasks. Now she was able to move it into any position. It was the trauma that was locked in her tissues and when it came out with the shivering, her arm was able to move.

I sent her back to her deep tissue therapist as yes, she needed more soft tissue work but she now had the movement of her arm back and was connected to the cause of the long-standing problem.

"Seeing" without actually Seeing

A friend of mine, Maria, called me to come over to see if I could help with one of the children she cared for in her day care. She is a wonderful caregiver and rarely had trouble with a child. This one would not stop crying. When she opened the door, she was holding him. I think he was about 3 or 4 years old. I sensed energy on the top of his forehead and asked her if I could touch him and she agreed. As soon as I put my index finger on that 'spot', he stopped crying. No, I don't know what I did, nor did I see anything with my eyes. It was a knowing gut feeling, and it worked. She never had that problem with him again.

I was at a meeting with my friend Leta. She had a beautiful crystal on a chain around her neck. Immediately my attention was dramatically pulled to that crystal, not knowing why. I asked her if I could just put my finger on it. She agreed and as I put my finger on the crystal all this energy came out, taking both our breaths away. It wasn't even a moment before the energy was cleared. Again, I was Seeing without visually seeing. A gut sense told me there was something there to be attended to.

Body Memory

Here is another example of a session that relieved a body memory using the body to recall the sensations at that time. Carlie is a sweet woman in her 40's and asked in group one evening if she could be the demo subject for a trauma resolution session.. She explained her breasts had been touched inappropriately by an older man and when telling her parents, they did not believe her. They sent her back to clean his house again. She was 15 years old at the time.

In later visits, she was able to avoid him, but the initial trauma was hiding in her body. She had gone to many therapists to clear it and with no results. That evening in class as she agreed to be an example for the lesson, I asked her to drop down into her heart and let the sensation of her body become very real. Getting grounded like this gets the client in touch with their body so then the body can work the session with no repeat trauma. It is important not to let the client get hysterical and dissociate again. Keeping eye contact is necessary. If you are grounded, eye contact with her will keep her grounded as well.

I asked her to slowly tell me the story of what happened and if and when she felt any kind of a body sensation, we would stop and experience that. As we went through the sensations she mentioned she did not remember the entire experience. She could not remember what happened after he first touched her. (Dissociation took place) This release took about 20 minutes, slowly letting the body sense each memory and then when it resolved, she continued her story, again releasing trauma. She had a few tears, and I had to ask her to open her eyes and look at me and she then calmed down. (She self-regulated through connecting with me being grounded) The various sensations she was having in her breasts were very uncomfortable, but she was staying grounded and getting through the memory.

When we were finally finished, she looked up at everyone and said how differently she felt. She then remembered the rest of the experience…she had actually pushed his hand away and left his house. She empowered herself back then and didn't remember as the trauma was too difficult for her young girl to handle. She also suddenly realized that it had never dawned on her why, when making love, she never liked her breasts touched!

Here is a great example of the subconscious mind protecting the body at all costs. Since she had not resolved the trauma, the SC was keeping her breasts from being touched even in a loving space with her husband, but she was not even conscious she was doing this! It was like a light came on in her mind. Her eyes got big, and she was even more surprised than all of us in the class.

The SC mind will protect you at all costs. Once the SC mind 'got' that the danger/trauma was resolved, there was no reason to stop the touching that no longer held the trauma.

Some of the other chapters will give more clarity and detail so the reader will understand how to do this process. It is important to not jump the gun and start playing with this right away. You can drive the trauma deeper if you don't know how to work with it.

Getting SC Help

The SC mind can be very helpful. I tripped onto this accidentally. My mind was used to being peaceful for sometimes hours at a time as I did massage or craniosacral therapy. By being peaceful and focused for so much of my daily life, I became experienced in meditation even though I didn't have a purposeful daily practice. I also learned that if I am relaxed and have both hemispheres of my brain working, (*Hemispheric Kinesiology* www.hksuccess.net) the answer to any mental question will show up shortly as a thought.

The more you train your SC mind to do this, the more easily and quickly you will have a response. It especially works for lost items. For instance, I might mentally say "where did I leave those keys", I put my thumb and index finger together to get both hemispheres working and then relax. Then I let it go and get about my day. Within a few minutes, the answer pops into my head. You must be relaxed, pose the question, be grounded, (thumb and index fingers together helps) and wait.

Jewelry Holding Pain Memory

After my mother died I had a piece of her jewelry nearby in my bedroom. Remember, I am an empath, I feel pain and emotions of other people. I woke up in great pain in the middle of my gut and felt as if I could not get any air into my lungs. I really felt I was going to die. As I got up and went to sit on a chair in the room, trying to breathe, I said in my mind, where is this coming from? An immediate answer came to mind, "This was your mother's pain" and with that it completely and instantly went away. My mother had lung and pancreas cancer therefore the great pain and difficulty in breathing. I then moved her jewelry out of my bedroom. I wonder what would have happened if I had just kept feeling the pain and not asked. I might have made it "mine" and suffered a great deal as it continued.

If I had not been grounded and aware of this sudden unexplainable feeling of pain I might have panicked. I asked the question of my SC mind and got an answer that allowed me to know what the problem was. My practice in being conscious and feeling my body was instantly a resource for me to self-regulate and ask the question. Imagine if I hadn't had that skill. I might have wanted to go to the ER and they would have found nothing, yet I still would have had the pain.

From Dragons To Butterflies –
Trauma Resolution & Morphic Field Energy Healing

Chapter 3

THE BRAIN AND THE MIND

Even though I have gone over so much of the brain/mind connection there are a few tidbits that I think will be important and a nice addition here, even if repeated.

I hope you now understand that the brain and the mind, however intertwined, are really two different experiences. The brain is the physical part of our human incarnation within this physical time. The brain allows us to automatically breathe and live and move in the physical world. Our mind uses our brain also to function in the world.

However, the mind is special. It can travel anywhere and see things that are beyond imagination as well as use intention to heal or change a situation. Someone in a coma may have an active mind, somewhere in some dimension. I remember the true story of a paralyzed soldier from world war one who mentally was yelling out for help, but of course no one could hear him. For years he lay trapped in his body, all the while his mind was trying to reach out. His mind was active but his body was not. Finally, a nurse realized he was 'in there' and he was acknowledged and worked with until he could speak and move. While in a coma, I would imagine the mind is in another dimension but who is to say for sure?

If the mind has been damaged from trauma, it can create an energy field compromising the brain and cause depression or even physical damage to parts of the brain as in some mental disorders. As I have explained before about the left and right hemispheres of the brain, early childhood trauma or any trauma in life can shut down one or both sides of the brain therefore stopping the mind's use of these very important parts of our brain for daily life.

A simple example would be parents who made their child do everything perfectly. Each time their bed was made it was inspected for wrinkles. Clothes had to be hung in the closet according to color and style. The child had no play time, only work and each day was filled with chores. I had a client who as a child was punished for laughing or smiling. There was absolutely no joy in the home. She ended up being very left brained as no right brain development was encouraged in her family. She was joyless and a workaholic.

The extreme of this is having no boundaries. That does not allow the left side of the brain to develop. A parenting example of no boundaries is a parent who does not stay consistent with parenting behavior. Spanking the child one minute and letting them get away with something similar in the next moment. (I do not agree with corporal punishment, BTW.)

Being lax about encouraging the child to have personal care of their bodies or getting to school on time or no rules at all in the household is an example of no boundaries that leaves the child out in the cold so to speak about living life in a supportive way. If they have no support as a child, it is likely they will not know how to support themselves or anyone else in the future.

This of course affects the mind/brain connection to life. Parts of the physical expression have become unavailable so thought patterns will fit whatever habits were learned during this early formative time. What the mind doesn't know, it won't know unless some intervention is applied. Once the child grows to adulthood the patterns are set. If the adult is exposed to an understanding of being able to have a better life, motivation then becomes the impetus to change.

It would even be possible for the child to develop a dissociative disorder where both sides of the brain were weakly active so most of the time the child or adult would dissociate or not be embodied. This results in a tragic loss of embodied life experience.

Multiple Personalities

First I would like to explain the difference between Persona and Personality. We "put on" different personas when we are in different social situations. If you are a judge trying a case, you will not express casual laid back conversational tones to the courtroom. Your persona would be that of a judge holding court. If you are a mother who is sweet and cuddly to her children, you would not do that in a social situation with friends. Your persona would be more reserved yet also having a good time.

Personalities, as related to MPD's, hold a different context. These personalities have developed into people who are held in the mind who emerge from the core personality as the occasion fits, or sometimes as the occasion does not fit. They have great differences in belief systems, size, culture, even ethnicity. When someone is traumatized enough to develop an MPD, the mind will choose just the right person type to help in that situation.

For more severe trauma leading to dissociation, multiple personalities can develop. I know I was fascinated as a child with a movie that came out, *The Three Faces of Eve* which of course was about a multiple personality disorder. Then *Sybil* came out when I was a bit older and I could not watch that enough times either.

I have always been fascinated with behavior and as a child was determined to figure out the why's of behaviors that I was noticing in the adults around me. When you understand that multiples are a survival mechanism for the child or adult to have 'help' facing trauma, you can be compassionate about the situation. Dealing with the aftereffects of such dissociation can cause serious difficulties in the adult life. (See *A Fractured Mind* by Robert B. Oxnam)

There are variations on the severity of what can happen as you might well imagine. I had a client years ago who called me in a panic. She had arrived home from shopping and noticed she was unloading many items she didn't remember purchasing. It turned out she had shop lifted without remembering. She asked me for a session and as we began I tuned into all her personalities. (She had known she was an MPD for years) I 'asked' all of them who was responsible, and Marilyn stepped forward (all of this is mentally viewed by me) and admitted stealing. Her reason was that she never got to 'come out and play' and needed exposure time. We 3 made an agreement that she could step forward during my client's classes that she taught if Marilyn would only stop stealing. It was resolved and no more trouble was experienced.

I gave extreme examples so you can extrapolate from there. If the left side of the brain is not functioning very well, the adult might end up with addictions since the left brain allows us to set boundaries. If the right brain was never developed well, the adult will most likely turn out to be a task master, a workaholic and a demanding autocratic parent. Remember the left and right brains need to be both on board and working together for life to be balanced. Imagine that one is hot water and one is cold and when both working, you have nice lukewarm water that is comfortable and a balanced personality.

Dissociative behavior (with or without MPD) can cause all kinds of havoc in one's life. It can be from minor to major. If we only dissociate around food we will either over or under eat or even develop an eating disorder. Eating while dissociated can make you feel as if you have not eaten, then of course you want to eat again. You may also not feel that your stomach is full because you are not connected to your body. Perhaps the rest of your life is pretty good but when food shows up, anything goes. The same can be said for any addiction or compulsion.

I gave extreme examples so you can extrapolate from there. If the left side of the brain is not functioning very well, the adult might end up with addictions since the left brain allows us to set boundaries. If the right brain was never developed well, the adult will most likely turn out to be a task master, a workaholic and a demanding autocratic parent. Remember the left and right brains need to be both on board and working together for life to be balanced. Imagine that one is hot water and one is cold and when both working, you have nice lukewarm water that is comfortable.

Once this behavior is named or realized, the client can notice and make choices about what to do when in these difficult situations. At this point the client might have sought out help for the problem. If not, these simple steps will help. Again, I will emphasize about getting grounded, being very conscious, and being totally in your body. As author Jack Carson and teacher Dr. Richard Bartlett are known to say, "Just notice what you notice". Pay attention to body sensations. Take a deep breath and feel it go into your belly and chest, exhale and push all the air out slowly. How does that feel in your body? Are you noticing the immediate feeling of peace as you do a few breaths?

The next important step is to slow it down. That came from Dr. Michael Shea and trauma training. If we are zipping through our day or our food or whatever, we really aren't present for it. We are just getting through it to get to the next place. Slow down and enjoy each breath, each task, each moment of what you are doing. If it's a distasteful task, do it with more attention and presence and just that focus will train your body for more grounding and enjoyment in everyday life. You might even begin to enjoy a normally distasteful task since it will allow you time to slow down and breathe.

Learning to Choose

The big kahuna that I have found to be the key along with the above is Choice. When I finally quit smoking I had this idea to only smoke a cigarette when I could sit down, uninterrupted and pay total attention to each drag of the smoke, the smell, the taste in my mouth, how the ashes looked and especially smelled since I would put each cigarette out into an ashtray filled with butts and some water. After a day or so, the smell was abhorrent and the green nasties floating on top were very off-putting to say the least and being conscious of this habit really allowed me to put it down with no problem.

I learned that the desire (coming from the receptors in the brain) for a cigarette would pass in 5 minutes if I could hang on for that short time. I found it to be true. Once the receptors were no longer activated, I was home free and free from smoking. You can stop, slow down and use Choice in any situation in order to make a major or minor life change.

Changing Behavior

If you can totally be in Choice about doing something, you can be empowered and in Choice about stopping it anytime. I worked with a client, Sherry, with an eating disorder she had had for over 15 years. I asked that she consider choosing (not my demand or mandate) to eat consciously when she binged. Instead of eating so fast and getting it all down fast and then purging the next minute, I suggested she try enjoying each mouthful. Choose which piece to put on her fork and chew slowly, tasting every bite; connect with her stomach and feel how it was doing after each bite.

Then when she decided to purge, sit there and look at the food she let go. Allow all her senses to embrace the experience of putting all the food in and then throwing it out. I encouraged her to see her own body fluids and digestive juices which God had given her to digest and nourish her body floating there with her rejected food. Notice the smells and embrace all her senses in this experience.

After I gave her all the ideas of how to begin to change her habits, I encouraged her to continue her food issues, but with Consciousness and Choice. We kept in touch by phone weekly and as she told me her successes or failures, I continued to encourage her. She was determined to get through this and have normal eating habits.

We also had worked on a bit of trauma that she remembered as a young adult. Within a few weeks she called to let me know that she was eating normally. She had told her mother to have no sweets around so as not to tempt her and of course, she had to eat small amounts at first since her body was not used to food. She let me know she wanted to keep in touch now and then to say hi. She was on a new road in her life and her success was inspiring. She also told me she wants to finish getting her degree so she can help others with their food issues.

The key is to not make anything wrong. Understand that even alcoholism or eating disorders or other compulsions and addictions have proven to be a resource for the person to get through some trauma that was too big to face. Once they understand this, they can make a choice to face whatever problem or memory. The decision to release the so-called 'resource' they used to distract themselves from feeling whatever the trauma is will be up to each client.

It is important to note that in the case of fully formed multiple personalities not to encourage the client to let go or eradicate the multiples. Once the personalities are named and been made aware of, the key personality must set up the rules for the other side personalities.

Dropping down into your heart is one process that I start every class and session with. I know that if I am in my head or brain I will try my best to figure something out. If I am in my heart I then just 'know' and feel the information much more clearly. "Figuring something out" is not for working on yourself or your clients. Being in your head too much is more of a roadblock than any benefit. To nurture compassion, intuition and gut knowing the heart space is the place to be.

I like to play with some visuals as these get us into our right brain or right mind. I sometimes picture a pink parachute to gently waft me down into my physical and etheric heart space. Sometimes my heart looks like a lotus flower or a princess castle or a beautiful waterfall. Please get the picture that I imagine getting myself down to this heart space and all the while being grounded, breathing, and feeling my body all the way out to my skin.

Being in your heart will bring you the ability to tune into someone or something else. This is where the peaceful mind comes to the fore to assist you in traveling mentally or knowing or sensing. It is all expansive and has no limits. You can only get there with deep relaxation. If you need a musical background to focus better, then let that be part of your dropping down into your heart space.

The Institute of HeartMath (Search Google for wonderful aids in assisting to relieve stress) offers classes and other equipment to enable one to get to that peaceful place. It is a great website to check out.

The Buddhists are known for their meditation practices. Whenever I mention meditation the other person invariably says, "Oh, I never could focus on just one thing for even a minute." What they don't know is that it is not necessary to do that so perfectly.

Meditation

Two types of meditation will be mentioned here. Both are easy once you learn to be settled in your body and be peaceful into your heart space.

One is a breathing meditation that was explained again in Jack Carson's last Gremlin book on Mastering your Gremlin. The details he writes are worth researching. The method is to assume a comfortable seated position and simply dropping down (into your heart) and consciously breathing, one breath at a time, in and out fully.

When you take a breath in you will want to breathe into your belly and breathing out will collapse the belly to expel all the air out. Counting one two, one two with each two breaths in and out can help concentrate your focus. If you begin to think of something else and lose the focus of the slow breathing in and out, just return to the in-breath and out-breath, fully and slowly. You may close your eyes or just focus distantly or have a beautiful picture in front of where you are sitting. Maintain this as long as you can and you will find the more you do it, the longer you will want to stay in breathing meditation.

The next meditation is part of a Witnessing procedure I will be explaining later. It is simply again being seated, dropping down, breathing deeply, and eyes closed and *just noticing* your thoughts. That means no thinking *about* your thoughts, just noticing with NO judgment at all, what each thought is as it comes up in your mind. If you again get off track, just return to noticing your thoughts, not thinking about thoughts but becoming the observer of what you are thinking in a non-judgmental way.

You may find one or the other most pleasurable after a time and choose to have longer periods of this stress reducing mental state.

Witnessing & Observing

There is another mental process called *Witnessing* that I stumbled into as a young mom before I even knew it was a taught procedure. It was usual for me to have a rough day with my 3 children. I had little to no parenting skills and somewhere in my gut I knew I was missing something important. It dawned on me one night lying awake in bed to 'look' at my day as if I was watching a movie. Since I have always been so visual, it was easy. I imagined sitting in front of a mirror and watching my day as if it was a movie rerun. I had it start in the morning and mentally requested highlights.

As I watched my mental movie, I realized I had no judgment of my day or my actions and could clearly see where improvement could be made and set about planning to do better the next day. Years later in Dr. Shea's craniosacral classes we were taught to Witness. I also use it in my classes as a mental exercise of getting in and out of our bodies consciously and looking at ourselves from the other side of the room. This can be a cool experience of seeing yourself from another perspective. If you can manage split attention you can be in a conversation with someone and 'see' how you are interacting by Witnessing the both of you interacting as you mentally step away to view the dynamic.

There is another facet to Witnessing that can bring you more deeply into a daily routine of peacefulness and understanding. Just imagine for a moment that you are watching your husband complain heartily about his job. You can choose to jump into the fray and become part of his upset, or you can observe what is going on and just be present for his dilemma. For you to maintain being grounded and observing can help to lower the elevated emotional state of what is happening in front of you.

Yes, you can be compassionate and understanding but not raising your emotional level to his will even sometimes allow thoughts of a resolution to show up. What I am describing is being an Observer of events outside of you rather than bringing it into your body or jumping into the pool of emotions played out before you. (See Lynne Forrest, *Guiding Principles for Life Beyond Victim Consciousness*)

Listening & Interacting

Of course, you won't want to stand there like a statue so interactions like naming what you see can be helpful. For instance, "I can understand why you are so upset" or "I see that you are really angry", naming the emotion you see before you. Also saying things like "I understand" or "I hear you" or variations of that such as "I really understand what you are saying" can be supportive and calming. (See *How to Talk so Kids will Listen and Listen so Kids will Talk)*

All the while this is happening you will want to stay grounded and centered, holding heart space to support the situation into resolution or at least peace.

Grounding Exercise

When I teach how to be grounded in the body I use an exercise somewhat like biceps curls using weights because the result strengthens one to notice when they might be in or out of their body. It goes like this. My client is there seated in front of me. I direct them to be grounded, feeling all parts of their body connecting to floor, seat, skin, breath, etc. When I see they are quite grounded and in their body, I ask them to think about their kitchen or driving in their car or whatever comes to mind. I give them a minute to experience this and then ask they come back into their body grounding again.

They practice going in and out and the more they do it, the more their body 'gets' the difference between being present in their body or dissociated.

The other day I had a doctor's appointment for a checkup. I accidentally took a different road (which turned out better than the original directions) and also had a rental car that I didn't like driving AND was stressed not remembering the roads I had to turn onto to get to the appointment and on time. I was aware all the time of my upset yet reassuring myself that I could relax as best as I could and just keep going and being present.

By the time I arrived I knew my blood pressure was up. Now that I was at the office I knew that I could get it lowered within a few minutes of dropping down into my heart, breathing calmly and fully and relaxing. It was only a matter of minutes that I affected calm and peace.

As a follow up chuckle, I ended up on a different road going home in the same rental car. The road was a narrow 2 lane road, and the distinct gut feeling was that I was going south and needed to go north. (I missed my compass in my other car) Indeed that was true. I stayed calm the whole time and just enjoyed the ride and the scenery. After I turned around to go the correct way, it became an amusing experience for me. My husband and I had a great laugh later about my driving difficulties.

This is to demonstrate that there will be sometimes when you are 9-1-1 but just know you have the tools to calm yourself within minutes. You might even find it amusing later so that later events like this can be treated more lightly instead of with rising drama and difficulty.

Emotional Intelligence by Daniel Goleman is a great book that offers insights to behavior that can only improve your life. I hope you can see that I may not have graduated with a psychology degree but life's experience and my unflagging interest in discovery of better ways to live led me to books, experiences, people, and assistance in my own life that brings me to these pages to share with you. If you don't like to read, get the books on tape or with some Kindles, Notebooks or iPad you can download a book and it will be read to you. There is no shame in not knowing it all.

Hypochondria

As is mentioned in the trauma chapter, symptoms of all sorts can show up in the body without the person realizing it is from a long-ago bad experience. A case in point is my mother who was incested by her stepfather. I was able later to name what was going on for her, hypochondriasis. When she found out what a migraine headache was, she 'had' one the next week and the week after that and so on. Eventually I found myself acting out with these symptoms and yes, the HC (hypochondriac) does indeed feel these things.

According to Richard J. Davidson, Ph.D. (*The Emotional Life of Your Brain*) the insula cortex in the brain can be larger than most of the population therefore being more reactive to body sensations and thus the thought that 'something must be terribly wrong with me' syndrome.

There are separate and sometimes a combined set of problems occurring on deep mental and physical levels. The body, holding the trauma, will show up with all sorts of symptoms, pain, illness, auto immune diseases, and so on. In addition to the hypersensitivity of the insula cortex in the brain, anything going on in the body will be felt more distinctly and therefore raise up a red flag of fear. The hypochondriac will also look for ways to get attention (Munchausen Syndrome). Many times their life history is loaded with fearful experiences, so the body looks for more ways to substantiate that fear, in other words, a life-is-scary-and-dangerous belief system. Reading symptoms online or in medical books will convince the reader that this disease or disorder is something they now have. I have referenced *Phantom Illness* in the recommended book section at the end of this book.

Panic Attacks

One more disconcerting symptom can show up if one is highly self-aware, panic attacks. Yes, you can *feel* or *sense* too much to the point that the SC mind sends more fear messages which develop into panic attacks. A scenario could be - you hear symptoms on a TV show. You start thinking about those symptoms and can totally relate those to your own body experience. When the diagnoses comes out over the TV program, it is dire, and you are then sure that you will die soon from this horrible disease. Your palms get sweaty, it's hard to breathe, and your heart feels like it will come out of your chest. This mentally constructed fear experience, not understood for what it really is, will demonstrate to the person that, "Oh No!" and then the activation escalates to fears beyond fears, increasing the panic attack and more mental fears.

How can you deal with this? The very best thing is to turn off the TV show and go to a quiet place and calm down, get grounded, and know that you are safe. By allowing these reactions you are terrifying your SC mind which is totally convinced that you are about to die in the next few minutes. Breathe deeply into your belly, let yourself know you are safe, and this is only an activation, not a real experience. Note the sensations in your body and watch them dissipate. Do not close your eyes but focus on everything in the room and notice that it is just there, in front of you, and there is nothing to fear.

The true difficulty is that what one believes can show up in the physical. Remember, we do create our own reality, and the mind is a powerful creator. Holding onto these scary thoughts and made-up stories can possibly allow them to truly show up in your life. A great quote I once heard is: "Worry is a misuse of the imagination". Usually once the activation calms down the person will continue to worry and think about these dire possibilities. Let it go, know you are safe, and allow your mind to be at peace.

Further on in the book, I explain how to use EFT (Emotional Freedom Technique) to relax the body away from panic attacks and self-regulate.

Chapter 4

BELIEF SYSTEMS

According to Rick Carson who has authored a few excellent books, among them, Taming Your Gremlin and A Master Class in Gremlin Taming, "…beliefs, even the noblest of them, are just opinions you have developed loyalty to."

I couldn't have said it better! And the best yet is my recommendation that you read and work with all of his books. They are fun and entertaining and cut to the core of cleaning out your thoughts, especially the ones that do not serve you or especially the thoughts that want to bring you down.

Or in short form, Belief Systems can be so much BS. I think you get my drift. And at the same time, what you believe about the world will almost always manifest into real experiences. If you feel all women are manipulative and conniving, guess what. Those types will be drawn to you because of your belief. If the world is a painful and difficult place in your mind, it will again show up in the outside world as painful and difficult. As you will see below, we really shortchange ourselves in our lives if we have such short sided beliefs.

Bruce Lipton's book, *The Biology of Belief* demonstrates that what we believe can end up in our body biology. If you know that your family has a history of cancer, you can bring it on just by believing that of course, you would have it too. Cancer can be a strong morphic field and can be enhanced with your belief to bring it into your reality.

Our belief systems can limit our view of life, literally. If you had some bad experiences with men as a child you may believe that all men are trouble. I call this seeing life through a veil colored with past experiences. An excellent example of this is a poem I loved from childhood by Leroy F. Jackson called *Grampa Dropped His Glasses:*

Grampa dropped his glasses once
In a pot of dye,
And when he put them on again
He saw a purple sky.
Purple fires were rising up
From a purple hill,
Men were grinding purple cider
at a purple mill.

Purple Adeline was playing
With a purple doll;
Little purple dragon flies
Were crawling up the wall.
And at the supper-table
He got crazy as a loon
From eating purple apple dumplings
With a purple spoon.

What are your glasses colored with? It could be that
"no matter what, I can never make enough money" or
"no one ever pays attention to me" or "each time I walk
into a room, everyone stares at me" and so on. Be
mindful of repeating phrases that go round and round
in your head, they just may be some old colored glasses
from a belief system that no longer serves you.

Perspective is an important consideration. When
visiting an art gallery, I have witnessed the observers
moving around to look at different angles of the piece
of art. They are experiencing the full effect of the piece
in all perspectives. Conversely, if you have one belief
or singular perspective about one thing without
entertaining the possibility of other ideas or
perspectives of that thing or idea, you will be locked
into a mental room with your opinion and the lights off.
You will literally be in the dark with your eyes closed

to any other possibility. Even if you know other perspectives or ideas of whatever is the subject it does not mean you need to adopt them. After all, knowledge is power. Open your mind, take off your purple glasses, and SEE the world and all that is in it. Understanding what other people see about any subject or thing can broaden your own scope of understanding and add more compassion to your life.

Sometime in the 70's I was a student of a brilliant teacher. She was weird by most standards being a powerful empath, visionary, and a traveler into other dimensions. Sometimes we would be on the phone for hours where I would learn about all things that are metaphysical and energetic. A short time after we began our work I had an unusual dream which I recounted to her.

Informative Dreaming

I was looking in my bathroom mirror and I was wearing a black and white tuxedo. It was so small on me that it looked like a doll suit. I kept pulling at it, trying my best to make it fit. As I looked into the mirror beyond my absurd reflection, I saw that the back of my house was gone and there were no walls. Everything was open. I knew this was a good puzzle for my teacher to help me understand what this dream might mean in my life.

It turned out that the black and white too-small tux was my belief system. It didn't fit anymore. And yes, my catholic upbringing along with the prejudice my parents had passed to me had allowed my world to become black and white. It either was or it wasn't. In other words, there was no space in my mind to allow for mistakes and the stumbling blocks of life's lessons. I realized I had been critical of everything and everyone, especially myself.

The back of the house being gone signified my vulnerability, or rather my feeling vulnerable. When one's belief system is challenged and everything you ever thought or believed in is gone, it shakes you to your core. I felt open for attack, flailing about in a dark space of my mind's eye. I had no anchors to tell me that my world was predictable and safe.

The final part of my dream was the cherry on the cake. The scene switched to the outdoors where I was looking at the beginning bare bones of a towering structure in the process of being built. The girders and struts were not connected correctly. I could see sagging places and some areas where nothing was connected at all. This lovely mess signified the collapse of the inner structure of that belief system, finally beginning to fall apart. It was time to rebuild.

Everything I had done up to this point was to figure out why life was the way it was and why what happened to me early on might have been a benefit or a curse. I was filled for years with anger and frustration that everything I had expected in life was nowhere to be seen.

If you can finally 'get' that your life's path is all about finding Truth and Understanding you are ahead of the game. Compassion is the big one I still wrestle with having come from such a judgmental background. If I leave it as "they who raised me screwed me up" and don't take responsibility for my own life, I will forever be caught in the swill of the sewer of life as I circle the drain. Blaming the past on your current problems only makes you a victim. When you are a victim and play the blame game, you are stuck in your own cesspool. Never taking responsibility for staying stuck will keep you there indefinitely.

The Blame Game

Maybe I can clarify that with an example. Let us say that your dad was an alcoholic and your mother was abusive. Your friend asks you why you are complaining that you are again stuck in a dead-end job. Your answer is that you just can't seem to get ahead. Your mother and father stunted your emotional and mental growth and there is nothing you can do about it. You finish that with a glum and hapless look in your face, resigned to your fate.

That is not taking responsibility for your own life and moving forward. Yes, you might need some kind of therapy but if you constantly remind yourself and everyone around you what bad things happened to you in the past, your SC mind is reliving that experience and driving it more deeply into your psyche. The "It wasn't my fault" mantra will only keep you mired in your misery.

You can see this is another false belief system, locking you into a limited life of "poor me-ness" that you are continuing to cultivate. If I mention what happened to me in my childhood, I look for body sensations. If I get activated, I know that I have more that is coming up that I need to clear. Of course, there is the dissociated conversation where you are not feeling anything and just prattling on and on daily about your sad course in life because you were tormented as a child (or whatever). I am not diminishing the awfulness of that, just mentioning that it may be time to move on and get the trauma resolved so it just becomes a memory with no emotions attached to it.

I remember the name of a book; I think it was *"A Boy Named It"*. His (crazy dissociated) mother kept him locked in a basement and plied the cruelest of tortures on him. Thankfully he was eventually discovered and rescued and later became a public speaker to let people know that what happens to you does not necessarily define you.

Sometimes one cannot really name and remember the event that gets triggered. All they know is that when they see or hear or feel something, a big abreaction comes up. The initial incident might have been pre-verbal, in other words, as an infant or young child. In this case, there are no words at all to say, only body sensation. Ask the client to slow down, breathe, get body connected (grounded) and imagine going to that experience of "something happening" getting them activated over and over again.

At this point, the intelligence of the body just 'goes there'. The client may not see or remember anything but the body starts with its abreaction because it has been asked it to go to the place that is a problem. One never even needs to know what it was, just that it will finally get integrated into the body where it belongs and release the emotional flurry that had accompanied it. This is implicit memory, pre-verbal, and can be released with the trauma resolution work.

How does that work? Remember I said the *body remembers*. Picture this. Some trauma, big or small, occurs and the client dissociates. Imagine a cloud of that incident hanging outside of the client; BECAUSE THE BODY has not yet integrated the entire experience. Every time something comes up that is even remotely like the event, the trauma cloud knocks on the door of the client's body, wanting to be acknowledged and to be invited in. Those would be the body sensations we feel when our body is remembering unresolved issues not yet integrated.

There will be more about this in the trauma section. If the client believes (<u>believes</u> being the operative word) that this procedure wouldn't work for them and that it is all hogwash, it won't work. This would be a good demonstration of a belief system. In that case, doing the work will not bring the client to resolution.

Perhaps they are firm in the belief that walking a labyrinth while doing a chant will help or taking a trek into the Appalachian Trail would be an excellent way to clear. For the client to be so invested in these experiences will most likely result in a beneficial outcome.

Belief Systems: Powerful!

Be mindful of them and know that one way is not the only way someone can go. In addition, if they are staunch in holding onto their issues, not believing them to be life limiting, back off and let go. It is not your day, or any day your day, to *fix* them. Really, not one of us is 'broken' so that we need to be fixed. Sometimes we just feel that way.

If you set yourself up to seek out those who 'need' your help, be careful. It might turn into a rescuing endeavor and you then become part of the Triangle; Victim, Rescuer, and Persecutor. (Lynn Forrest)

How do we get in touch with our belief systems to see if they are working for us? You can do a few things to make your discovery.

Getting to the Core BS

You can write down a list using your SC mind starting with:

Relax, have paper and pen ready. Get grounded. Notice how your body feels, sense the skin of your body connecting with the outside world and then ask a question, such as: "The first thing in my belief system is that I believe…" and write down the very first thing that comes to mind, not matter how it sounds. Stay grounded. Continue with, "the next thing I believe is…" and write that down.

Or if you are more of a visual person, get grounded again and ask your mind to show you a picture of what you believe to be true. Playing around with a few things will get you some information and at that point, you can decide to evaluate what works for you and what doesn't.

For instance, I was raised Catholic. I was taught in no uncertain terms that eating meat on Friday was a sin.

Some years later, that 'sin' was revoked whereby we all could eat meat on Friday, and I was wondering, as the child that I was, all those people suffered in purgatory for something that wasn't a sin anyway? Or did the Pope say it wasn't a sin now and that those other people who suffered in purgatory all that time was a waste? It didn't make sense to me. I was also taught that it was a sin to French kiss unless you were in the goal of having a baby. I don't know what the Catholic Church teaches now but as I remember the catechism class where this was taught, it did not make sense. AND at the same time, if you believe this and like the practice of it, then that belief system supports you and you need to keep the practice.

My grandmother was superstitious. If she even thought about walking under a ladder she was freaked out. The irony is that if you do walk under a ladder, it really could fall on you or something up there on one of the rungs could get knocked over and hit you on the head. It is a safety issue, not a superstition The same with breaking a mirror, it is not 7 years bad luck but certainly breaking a mirror can cause a problem with glass, injury, and a big mess.

Remember as a child this saying, "step on a crack, break your mother's back"? There is a Jack Nicolson movie (*As Good As It Gets*) where he plays an OCD guy, very left brained, and when he goes out walking anywhere, he cannot step on any cracks. They never said in the movie what he thought would happen, just that he would stop moving at all if there were only cracks ahead where he could not step. This character had to have his silverware brought from home if he went to a restaurant. When he came home, he went through a series of door locks that would dizzy anyone trying to figure out his sequence. It was a belief system that caused him to go to such extensive degrees of behavior and limited his life daily.

All of this is unconscious, or from the SC mind as we have mentioned. There is a feeling of unrest, not being safe, if these patterns are not kept sacred and acted on. You the reader may look at these habits and say in your mind, how silly. It is not silly to these people who have this belief system about how to BE in the world. Usually there is some kind of trauma lurking in their psyche, their SC mind, that has them create such seeming safety measures.

Being in your heart space allows compassion and understanding to come in. Understanding that in some place in their mind, this calms their angst. These actions make their world tolerable. This belief system is comforting for them and forcing them to stop is not healthy at all.

We can only change ourselves, not others. We can easily change our <u>reaction</u> to others much more easily than trying to make a change in them. If they seek out help as a client, then you both work towards less trying behavior. Aside from that, if you are in a personal relationship with someone, deal with it and do your best to have compassion. Change your reaction to whatever is going on and do your best to be in your heart space. There are always more options that come with compassion.

I will finish with another excellent quote in Jack Carson's *Master Class in Gremlin Taming*:

"Mahatma Gandhi had more than one poignant comment on this theme" "The Golden Rule of conduct…is mutual toleration, seeing that we will never all think alike and we shall see Truth in fragments and from different angles of vision."

The Yes-But Game

This is where we can fool ourselves and others into thinking that we are striving to be better. So we ask for advice from friends or therapists and either in our minds or out loud we say "Yes, But I already tried that and it didn't work" or "Yes, But I can't do that as I work so many hours" or "Yes, But my partner would never go for that" and so on. Be careful if you like using the Yes-But game. It is self-defeating. You may think you look like you are moving forward by trying to get help however you shoot it down immediately with the Yes-But game.

Being an Observer

Many years ago, I was in an emotionally abusive relationship. Later I realized it mirrored the one I had with my mother. I began to OBSERVE our interactions. I noticed he became more upset once a month more than usual. I noted that on a calendar. Soon I discovered that when the full moon was near to full, the worst of him started up. I began to do different things in the home that were more calming and moved toward giving him space and compassion.

Following that my experiment got really interesting. I noticed that when we would have some sort of argument, I would respond in the same way I did the last time. I began to categorize the similar spats by number. I would notice this was argument #37 so instead of responding and acting like I normally did, I changed behavior. He was unable to move at me in the same way because I changed the road signs. Wow, I was onto something here.

That really empowered me, and I made my daily experience all about observing and choosing my responses and actions. Once I felt empowered I was able to set boundaries. A caution here that is important to note. If you are in a physically abusive relationship, know that if you change behaviors it will threaten the perceived power the abuser thinks they have. You can be in danger here, so it is important to seek help from a therapist, a shelter for abused women, a family member and so on.

Here is what happens when a controller who has low self-esteem and only feels powerful if s/he can keep you trapped and isolated…if you change behavior and become empowered it is like this example. Imagine you are a passenger in a car with the person who has been violent with you. S/he has control of the vehicle no matter how fast it is going, and you are only the passenger with no power at all. If hypothetically you reach over and take the steering wheel off the column, the driver has no control and panics. An insecure physically abusive controller will always step up their abuse if they feel their power is being diminished in any way. This almost always results in increased violence.

In this case, please seek help and protection before anything else!! Remember, too, that someone who bullies another person usually has a very low self-esteem themselves from having been bullied themselves.

Relationships & Intimacy

Remember I mentioned that everyone is a mirror in our lives; it is especially true with long term relationships of marriage or partnership. I saw a movie recently called *The Center of the World.* It was about a wealthy young man who visited a bar where lap dancing was a favorite of his, with a favorite woman there. He offered to take her to Vegas for a weekend, and she agreed with boundaries set by her. When the boundaries were breached she lost her pretense of lap dancer sexy lady, she became a woman who desired a man. At that point, you could see she dissociated, she could not be present for intimacy, or better said, IN-TO-ME-SEE. Once her real self-showed up it was too threatening and she spaced out. He also was unfamiliar with intimacy so was confused when she backed out of what he felt was a choice and desire on her part. He did not understand that it was the intimacy they could not be safe with.

The final scene in the movie was him returning to the bar and paying her for the lap dances she did for him. In this, they were both happy with the arrangement, having seen that they were not candidates for intimacy. I doubt they each could name why they were more comfortable here but the gut feeling for both was such that this arena worked for them. The boundaries were set by the business she worked in, and he agreed to meet her there. She was safe with her persona of lap dancer with no 'threat' of being intimate with this man or any of her clients.

Being intimate means, you become vulnerable and have confidence and trust in the person you are choosing to be intimate with. They get to know about you, your thoughts, perhaps even some secrets you hold. Having sex with someone is not necessarily being intimate although that is sometimes what is thought of the words, 'we were intimate'. They may not have been intimate in the true sense of the word.

If your trust has been betrayed in your current or past relationships, it takes a while to trust again and with a great deal of work such as intimate conversation, boundaries, and agreements and possibly even couples therapy. Perhaps past trauma is the blockage in the current relationship and again, working with that will allow the doorway to trust to open again.

Remember it takes two to tango as the saying goes. Our partners are perfect mirrors for us to see ourselves reflected clearly. Look for patterns that are repeating in current relationships that may have occurred in either past relationships or childhood.

Chapter 5

MORPHIC AND ENERGY FIELDS

Sometime in the 1980s, we stumbled upon a book called A New Science of Life by Rupert Sheldrake. The entire book is quite heavy scientifically, but the specifics about the Morphic Field are priceless. Below is an excerpt from the 1995 revised edition, soft cover:

"All actual morphic units can be regarded as forms of energy. On one hand, their structures and patterns of activity depend on the morphogenetic fields with which they are associated, and under the influence of which they have come into being. On the other hand, their very existence and their ability to interact with other material systems are due to the energy bound within them. However, although these aspects of form and energy can be conceptually separated, they are always intertwined. No morphic unit can have energy without form, and no material form can exist without energy."

Before we proceed, let me say that energy is just energy. Once energy merges with creation or a being, whether it's a human, a tree, a car, or a bug, it is still the quantum particles that are put together to allow this to be in existence in the form it is in. Every physical experience exists because there is a morphic field holding this energetic form. If the tree contracts a tree disease, that disease has a morphic field and has become a part of the tree. The human being, with its body and organs, is also a morphic field, comprising individual morphic fields for each part of the body. Suppose an organ becomes ill for whatever reason. In that case, the illness itself is a morphic field within the morphic field of that organ.

These morphic fields can be felt distinctly by someone sensitive. We can all become aware of these fields by sensitizing our hands. If a human being becomes angry, the energy is flavored, so to speak, with anger. It is like making soup. You start with pure water and then add ingredients to make it taste different. Anger has a distinct flavor and can be sensed by someone attuned to such things.

I had a client who came to see me for craniosacral therapy because he was experiencing severe earwax problems. He had exhausted every other effort to stop the deluge of earwax blocking his hearing and causing pain. A bit into the session, I felt an intense anger coming from him. At the end of the session, I asked him if he had felt furious recently. He thought for a minute and said that he had indeed been a bit angry recently, especially with his wife. (Now here we are, Naming the Problem, and if you can Name it, you can Tame it)

He called me for another session the next week and commented that his wax production was way down. (Just mentioning his anger and having him own it, began to change the problem) At this session, I did a CST (craniosacral therapy) V spread, which energetically 'pushed' out towards the other ear, which was giving him a problem. At the end of the session, he asked if earwax had poured out of his ear and was on the table. He said he felt it come out. (He was feeling the energy) No, it didn't, but it sure seemed to clear the anger, as I never did see him again. I can only imagine that he has no more problems, as his anger and earwax issues are resolved.

Some of the trauma work I learned can be resolved just with a CST session, usually, though, there may be many needed. If you are not a CST practitioner, the trauma work recommended in this book is just as healing. Dr. Michael Shea (www.michaelsheateaching.com) teaches BioDynamic Craniosacral Therapy, an excellent way to resolve Trauma for body workers or those trained in CST.

Thoughts are things. If you put this book down for a minute, recall a joyful experience, and really immerse yourself in it, your body will sense the joy as if you are experiencing it again. (Here also, the SC mind is taking your body back to the experience) You can do the same with a not-so-good experience and notice the different body sensations you have with that memory.

In working with a client who has a trauma memory, you can ask them to remember that and run your sensitized hands slowly down the front of their body. You will "run" into a feeling of heaviness, buzzing, or even sticky thickness. If you feel around, you will find there is a boundary where you can push against this morphic field. Now imagine it as a bubble, like a balloon that you want to poke holes in. Spread your fingers wide and enter the field. If you are dropped down into your heart and grounded, you will know what to do from then on. (Also see the teaching chapters for more instruction)

You can even try this on yourself by thinking different thoughts and running your hands down your own field of energy.

As the field collapses, the person may experience a range of emotions and body sensations; again, it is essential to remain grounded during this process. Be mindful that it may even weaken their physical field, so make sure they are standing in front of a sofa or chair they can fall back into if needed.

Now, of course, the client can recreate the morphic field if they go back and think strongly of the Trauma, so it is necessary to let them know it is good to really let go of all of it once you have collapsed the field.

An energy field can be powerful, as I will explain. (For the most part, I will use the terms energy field and morphic field as interchangeable unless there is a specific need to name one more than the other.)

You may have been around people who are said to have charisma or power. Their standing in the world and in their body is healthy and strong. By the same token, a criminal who has committed heinous crimes will also have a strong field, but a sensitive person will not want to be near that field.

The following story will confirm the strength of a field, and in this case, one that has been in effect since childhood. My friend's husband called me on the phone to let me know he was taking her to the emergency room because she had such a horrendous headache that she could barely move her head. He asked if I could come over and see if I could do anything before he resorted to the ER experience.

Deanna

I arrived at their house, and when she opened the door to me, I knew not to touch her physically. That would be common sense, but at the same time, I got a distinct feeling of "do not touch". She walked into her kitchen and sat down on a chair. I dropped down into a grounded space in my heart and slowly began to move my hands around her head, about 5 or 6 inches away. When I got to the very back of her head, somewhere in the center, I felt a substantial, slender field. She yelped a bit, and I went around it and up on top of it again.

As I explained above, I went slowly around it and determined its shape. It felt like an ice pick, but when I

101

could 'see' it in my mind's eye, it looked like an icicle. I told her what I saw and that I was going to pull it out.

Important to know when working with a solid field like this, you will need to move slowly in pulling it out, always sensing that you still have it in your hand. I wrapped my hand around it and as I slowly pulled it out, making sure I still 'felt' it in my hand, she screamed in pain. My friend couldn't see me, as I was behind her, but she felt this come out of her head.

I then threw it on the floor, and she was immediately pain-free. At that moment, I saw an old hag standing there, grimacing at me. I described this hag to my friend, and she said the description fit perfectly of a woman from her childhood nightmares.

We had also determined previously that her mother might have had some being or entity on her as she had been abusive to her daughter off and on through her childhood. (We suspected that this hag had possessed her mother years before, maybe even as a child.) I mentally yelled to the hag to be gone, and she was. There was never a problem of this sort from the hag.

Here was this being who was a strong energy field, causing serious problems in the mother's life and the child's. You can banish these with words to that effect or even call-in angels to help. My favorite is Michael Archangel. I worked with him last night on a client, and when he entered the room, she could sense the power.

As his hand went to the area of distress on her body, she could feel it being placed inside her to facilitate the healing. Michael's energy is strong, clear, powerful, and loving. Not so with the hag. Big difference.

Katarina

I had had another client who kept saying that she felt she had some evil in her. (I could not see it at that time, so I thought she might have imagined it) Meanwhile, with each treatment, her spine became more painful. It didn't matter if I did energy work, CST, or Morphic Field Energy Therapy; she still had that evil feeling and more and more pain.

Her family roots came from a country that believed in and experienced possessions, as well as other occult and religious phenomena. (This is not uncommon in indigenous cultures, again, belief systems having valid physical implications) Finally, she called one day frantically asking if I could see her right away. She was so upset that she backed into a pole and then missed my building entirely, calling to ask why she couldn't get access. Within a few minutes of her coming into my office, I had her on the table to hold her feet to begin a CST session.

She began writhing in spasms and extreme pain. I just kept calm and held her feet. She started to scream in pain. At that point, I saw a woman who was similar to the hag I had seen in the previous experience.

What I heard from the woman was that my client had made a deal to gain power in another life in exchange for her soul. When the hag decided to claim her soul, she plunged a knife into her (previous life), thinking the soul would come out to her. (Which is precisely where Kat felt the pain.) Of course, this did not happen. The hag then took possession of the soul and claimed it for herself whenever it could be freed.

Because we banished her altogether in that session, there was never a chance for that to happen.

Later, my client and I discussed what she saw and heard, and it was a similar story to mine. We banished the one who had possessed her for two lifetimes (or possibly more), and she was free of pain as well as the feeling that she had something evil within her.

Consciousness and Energy

You can also add energy or a morphic field to anything. To charge a crystal with love and peace, hold it in your hand and visualize the idea of love and peace entering the crystal, then seal it in. Suppose you have mental images of babies, puppies, or a beautiful scene. In that case, it will help to form the experience to put into the crystal. This may be a bit silly but thank the crystal for holding this love and peace for you. EVERYTHING has a consciousness, and everything is conscious. If you expand your idea of consciousness, you will understand what I mean. There is a morphic field surrounding everything, as I mentioned earlier. There is also a consciousness within that field that makes it what it is. Energy, Consciousness, and Morphic Fields are all connected and conscious in their own experience of existence.

This example may help. My father gave me a huge oil painting of a clown wearing an old black top hat with a poodle on his lap. The clown looked a bit pensive, but otherwise it was just a beautiful painting. Well, my then-husband was creeped out by it. He mentioned every day that he didn't like the painting and wished we had never received it as a gift. After a few weeks, it began to feel creepy even to me. One of my sons had gotten up in the middle of the night and told me the next morning he thought he saw Abraham Lincoln in the hallway. (With the top hat, it did look that way) One night, my husband and I both sensed the clown's energy in our bedroom. At that time, I didn't know I could do anything with it, so I actually had a priest come to the house to do an exorcism. Yep, color us scared and crazy, but it was getting worse. After the priest's visit, we had no more heebie jeebies.

What I realized later is that my husband's fear had enlivened the energy in the painting. He made it real in his mind and gave the clown a strong enough morphic field to prowl our house. Even our neighbors noticed it one night when a cold air space moved from person to person and then disappeared. Believe me, it was weird. Now, not so much since I understand it.

You can know something, but until you understand how it works and how to work with it, it can become scary and unapproachable. Unfortunately, however, the more you know, the more you don't know. (The Socratic Irony) It is not essential to know, but to understand, and then only the things that interest me. One can learn Chinese, but to truly understand how the language evolved and why it features distinct intonations in speech or intricate characters, one must grasp not only the language but also the depth of the culture.

When I lived in Arizona, my husband and I visited a botanical garden with some friends for the day. I love sensing the energy fields of flowers and trees, so my husband and I were walking along with our palms open to the different plants, from cactus blooms to roses to trees. My friend's husband asked me what we were doing, and when I explained, he said he wanted to experience it too.

I sensitized his hands and brought him around to different plants. When we went to a Mimosa tree, we kept our hands 4 or 5 inches from the leaves. Still, the Mimosa 'felt' us there (this tree is also known as the sensitive plant, as its leaves will close if touched), and the leaves closed up. We had not touched the leaves, but our field of energy did. He was amazed. Then we put our hands on some trees.

One was very sad. I still don't know why. I brought him over to sense the tree, and he began to cry, a big, loud cry, quite upset, and being a guy, he was very embarrassed. I told him to take his hands off and move away. I grounded him, and the sadness went away. Some of these trees had been relocated from their native countries to Arizona, and this one came from Australia. Perhaps it missed its home. You might think that cannot be, but I have found it to be true in my many years of seeing and experiencing energy fields. Having done this work for so many years, I don't discount anything.

You may have heard that We Are All One and that there is Really No Time or Space; Everything is Happening Now. These two statements may be a bit much to understand or even believe; however, this too has been my experience.

I can be doing a phone session with someone, and if they get thirsty, I also get thirsty. Until they quench their thirst, I can drink water during the entire phone call, and I will not feel quenched. I am connected to this person, and they are connected to me. As an Empath, I am accustomed to feeling the pain and emotions of others. This gives me information, and I know not to own it. (see the chapter on psychic information)

During a phone session, I keep my hands free with a speakerphone and ask my client to do the same. It is essential that, as I work on them, they feel in their bodies what we are doing. That way is helpful as well. Suppose I am performing a craniosacral session and have my hands deeply into their cranium, working on a nerve or fluid. In that case, I want to ensure they are feeling OK while I am doing this.

I must be careful who I tell this to. They look at me like I have lost my mind. How can craniosacral therapy be performed over the phone? It is wild. We are all one. There is no (actual) time or space, only in the physical world, and THAT world is run energetically by morphic fields. I am connected to a morphic field with my client on the phone. That is how it works. If the client does not feel that I am working, I do not continue.

Phone Sessions

In one case, the client (Lisa) had kidney pain. Again, it is essential to realize that you may need to refer the client to a doctor, especially if the problem or pain returns. When I body scanned the woman, who was on the phone, I saw that it was the right kidney and it looked to be stuck, not moving. When I learned visceral manipulation, I was taught that every organ has movement, however slight, called motility. So I very gently reached my hands into the morphic field of the kidney and asked for it to come into my hands. I could feel the pressure when it did, and the client mentioned the sensation of what I was doing. I held the kidney through a series of still points (a craniosacral term denoting a sequence of quiet or releasing movements aimed at resolving reciprocal and balanced movements).

When the kidney felt as if it was balanced and moving nicely, I gently and slowly reinserted the morphic field of the kidney into the morphic field of the client. At that point, I could see the ureter was also stagnant and not moving, so I put my finger into the morphic area and worked with it until I saw another resolution. Meanwhile, the client reports that the pain has gone and she feels relief.

I could also see that the ureter had occluded the fallopian tube in that area and was a bit twisted with it. NOW, this is important. I examined both the energy fields and the physical aspects. So if the client had had an MRI and it had shown there was no ureter/fallopian tube involvement, that didn't mean there wasn't. Etherically, energetically, there was an energetic twist, and once that was freed, the client was quite relieved.

Yes, it helps to know anatomy and physiology. Depending on what your goals are as an energy worker, you will choose what you need to do and become a specialist in whatever area you come to love.

Another session with a client named Kyla was amazing.

Redoing a Birth

I was referred to a lovely young lady for phone sessions. (I later came to meet her in person when she came to my city) The problem was asthma, allergies so severe she was even allergic to water, and eczema all over her body, especially her face and arms. However, I didn't know about the eczema until after her sessions.

I knew some of her history from her mother, who told me she was the firstborn of three girls, and her birth was complicated. The mom had been in labor for many hours. The baby was stuck in the birth canal for a while, and when they finally got her out, her clavicle broke. No one knew that for three months, except for baby Kyla, who screamed constantly because of the pain.

As a young mother, you can imagine the dismay of a mother over a continually crying baby. When they finally discovered the break and repaired it, the baby calmed down, but the damage had already been done. She had implicit pre-verbal Trauma that showed up in her body.

The mom couldn't bond with her baby, which is usually accomplished by holding the baby with a calm nervous system, allowing the baby to model its own nervous system after the mother's peaceful state of mind. Hopefully, this is accomplished early on when the mom and baby are new to each other. You can imagine that the mother felt very insecure, unable to take better care of her baby. Of course, she would think it was her fault. The baby learned through this experience that the outside world is a painful place not to be trusted. The experience of three months of pain with no relief or comfort had been intolerable. As the child got older, symptoms began to manifest in a big way. She literally became allergic to the world, as you will soon see and understand.

Meanwhile, I am into my third phone session with her. The asthma is somewhat better, but the improvements were not what I had hoped for. (I would remove the morphic fields of her lungs and allow them to unwind; however, in each session, I found the lungs to be just as locked up.)

As a "coincidence", I had watched a TV show just before we spoke on the phone this one evening about a doctor in a third-world country assisting a midwife in the delivery of a local indigenous woman. The way they birthed babies was to have the mother stand with a cloth girth under the belly, legs spread, with the midwife waiting to catch the baby. In this show, the baby's shoulder was stuck in the birth canal. The medical doctor approached the local midwife (with permission) and reached in to turn the baby, allowing for a smooth delivery with no complications. Of course, during my session with my client, I hadn't thought of the show.

However, as we were talking during this third session (I don't remember about what specifically), she started crying. I asked why she was crying, and she answered that she didn't know. Suddenly, I realized we were in the middle of the body memory of her birth, as if we were there in that moment we were on the phone together. It had just come into my head that the reason I had finished watching that TV show before I called her back was to re-engage her birth trauma and allow it to be peaceful and safe.

When I realized she was crying because her body was remembering her birth (noticeable body sensations being reported, pain, and feelings of being trapped and difficulty breathing), I began to work with her to change that experience. I was also empathizing with some of her symptoms. I started using words to transform that birth into a blessed, loving, and safe experience. Step by step, I was able to get her to re-experience her birth as easy. I used visualizations along with the story I saw on TV of the baby that was delivered easily, with just a bit of intervention to release the shoulder.

I continued with pictures of her going home with her mother, being rocked, loved, and peaceful. No pain. This lasted approximately 30 minutes. She was still softly crying. She sounded relieved as she wept.

I knew we were finished and that she no longer needed to consult with me. Some months later, when she visited here and I finally met her in person, every symptom was gone. She had no more allergies, no skin problems, and no asthma. She is a beautiful young woman with lovely skin. I explained to her that her little self came into the world and, on some level, found it to be painful, irritating, and unloving. The outside world was toxic for her in her little mind. When that was shown to change, her body then accepted the world as a loving and safe place. This is a clear case of belief, past experience, and pre-verbal implicit Trauma leading to physical problems, and in her case, righteously so.

When it dawned on me that the TV show was so perfect that evening, I suggested to her that she go online and watch it to really sink in that she had a wonderful birth. She would be able to put herself in the same experience as that baby being born in the show to really confirm visually that she was able to rebirth and change her life. As you can see, there are no coincidences in life. Everything is in Divine order if we are just peaceful and know this to be true.

There is an excellent book by Carl Jung on synchronicity that provides valuable insights into "coincidences" like this one. Amazon.com will offer a full roster of his excellent books.

I have worked with two babies while they were still in utero. Before I even connected to the baby, I taught the mother how to do what we would be doing together. I did not want to be some strange energy that would penetrate the sacred space of the womb. If the mother and I did this together, the baby would know the energy of the mom, and together with her, my energy would be accepted as well.

Please be careful and considerate when tuning into or working with babies and children. Their energy fields are not yet wholly protected and formed, and a psychic intrusion turns out to be just that - an intrusion. You must ground and become totally peaceful before you even think of the mom or the baby. Move slowly and with reverence; never rush or exert energy during any session, especially with babies and children.

I recall a story that occurred during a craniosacral class. A therapist was working on another therapist on the massage table. In the CST practice, she had her hands under the woman on the table who had complained about pain in one of her vertebrae. The treating therapist decided she was going to 'fix' that vertebrae and mentally saw a hammer slamming into that vertebrae to move it back to normal. The moment she did that, the woman on the table let out a big yell of pain. NEVER even think such violent things when you are working with anyone. Their energy field will feel it even into the physical. As I mentioned earlier, be especially gentle with babies and children. Prepare yourself beforehand to ensure you are peaceful and in a good place. Then and only then will you be ready to touch mentally or physically a child or any client. Yes, I want to emphasize this point.

Everything is Conscious

I came to this conclusion after several years of working with the body, organs, and the mind. EVERYTHING is conscious on some level or other. Of course, the noticeable distinction is between humans and animals, however different their consciousness may be. According to Peter Tompkins and Christopher Bird in their book, The Secret Life of Plants (1973), their experiments showed that plants would respond to thought or spoken words. If the researchers were to discuss setting fire to the plants, there would be a strong reaction from the sensing equipment that had been hooked up to the plants. They even set up two plants connected and wired up so that if one plant were threatened in one way, the other plant would react, all the way across the country.

My experience with clients further substantiated that every cell in our body is not only aware but also quite conscious. I recall giving a reading for a woman, and at one point, we decided to see what a body scan revealed. When I reached the stomach, I felt an unpleasant sensation and, moreover, no digestion. I asked for confirmation, and she said this was an ongoing problem. I tuned into the most vocal organ I have since connected with. "It" told me (in so many thoughts and pictures) that she was eating the wrong food…that she ate what her husband did, lots of meat and starches. I told her that her stomach would prefer lighter foods, such as salads and soups, and small amounts of meat. At that statement, we both burst into tears! Her stomach was so relieved to hear about the food it wanted that it was overjoyed when she finally got it; we both felt the relief and ecstasy of "Finally!!! She hears me." We both knew that was why we cried with joy, and we just laughed at how her stomach could convey that joy to us so expertly.

I further deduced, after truly grasping and understanding that the Source or Universal Mind is everywhere, Breathing Itself and the Universe, and Living within everything, that we are all indeed one with Everything. (In Biodynamic CST, you can feel the in-breath and exhale of the Divine Consciousness as it breathes) Remember that when we die, the energy is not gone. The physical body that was us is transformed but not gone; it is just in a different form. If you have ever read "Freddie the Leaf," it is a children's story about death. Freddie knows fall and winter are coming, and he will be falling to the earth to be reabsorbed until he becomes part of the tree and will live again in another form.

This may be difficult for you to understand but do your best. Even though humans make cars and houses and so on, there is a consciousness in each inanimate object that makes it hold its form. And that consciousness is the Source or Universal Mind. If a table didn't have a table consciousness, it couldn't be a table. Source, Universal Mind, God is the Essence of Everything. Without it, we would not exist, nor would the universe or even your dining room table. Of course, the consciousness that exists in the table does not have the intelligence of a human or animal. Still, Source knows itself through this manifestation of itself, even in this inanimate object.

"There is this thing that makes things out of itself by becoming the thing it makes." Ernest Holmes

I offer you an exercise in working with your own cells and tissues. Suppose you practice meditation and know how to drop into Alpha or Theta brain waves. In that case, it is easy to communicate with your organs. I once thought my adrenals might need some help, and the visual I saw was two small triangles (the shape of the adrenals) with little arms and legs crawling through a waterless desert. Whoops, I wasn't drinking enough fluids! It certainly doesn't hurt to apologize to any organ for what you might think of as abuse. Of course, following that, you will do your best not to be abusive again.

If you experiment with this, do not be surprised if you sense movement in that organ! It happens to me every time. If you hear words in your mind, you may get a sentence or two. Or you see pictures like I did with my adrenal glands. Either way, practice and practice, and soon you will be able to scan and work with your own body, learning ways to help yourself that you never considered.

Working with Animals

A client, Sara, called me from Arizona because her cat was missing, and she wanted to know if I could help her find it. My gift is not necessarily to see things, but I can tune into animals. I got the name of every dog she had and finally found one who was intelligent enough to give me at least a sense of where Rosie the cat was.

This dog was part wolf, and the only information I could gather was that Rosie was somewhere high up. (I would mentally ask, and I would see the dog look up) I had Sara go through her attic, barn loft, and so on, but to no avail. A day later, a neighbor spotted Rosie on the top of a pole. It took a firefighter to get her down.

My friend Vickie is a dog person and has many. Whenever one dog is in trouble, she calls me to tune into one of the dogs who has passed long ago. What a brilliant animal. I get such information from Merlin that I could write it down as a conversation. Merlin will also stay with one of her sick animals or a new one for comfort. I am sure the living animal 'gets' that Merlin is there.

My friend Jackie called me about Bear, a Chihuahua Jack Russell mix that she rescued from the pound. Bear had been abused. The vet had noted that she had been kicked and kept in a cage so long her paws had become widely splayed in order not to fall through the spaces. At any rate, she had taken Bear to the vet, but the vet found nothing. I asked Jackie to bring Bear to my house, and as they were on their way, I sat down and started working on her. The side where the liver showed up brightly was where I knew I would work, so I began. By the time Jackie arrived with Bear, I was able to work physically as well. Bear jumped up onto the sofa with me, lay down on her back, and moved her leg so that her liver area was exposed.

Bear knew I had been working on her! I put one finger on that area and suddenly let out a sob or two, with tears, and knew I was done. (Animals and children are fast!) Bear jumped down, and as we both wanted to comfort her, she knew better. She was still "cooking" and needed not to be touched so she could finish. When we reached for her, she backed up. Duh. We finally got it and left her alone until she came to us, knowing all the Trauma was out of her body.

Friends of ours had one of the most intelligent dogs I had ever had the privilege of knowing. Annie was a shepherd mix, and Kathy had had Annie from a pup. She was well-disciplined and obedient within a minute of any command. Annie had been forbidden to go into their pool. Still, one day, when someone visited to swim, the woman dove into the pool and was swimming underwater. We think Annie thought there was a problem, so she dove in and swam under the woman, bringing her to the surface.

Another time, I was visiting, Kathy had told Annie to stay next to her, so Annie lay down and put her head on her paws. I thought I would try something, so I mentally whistled, and Annie immediately jumped up and came over to me, sitting down right in front of me as if to say, 'OK, I am here.' THAT was something! You can train a dog so well that they hear your thoughts. Kathy was astounded that Annie disobeyed her until I told her what I did.

Animals are WAY more intelligent than we think!!

Chapter 6

PSYCHIC STATES

Before you read on, I would like to clarify the ethics and boundaries surrounding psychic gifts. Please do not tune into anyone psychically unless you are asked and only in a place of assisting. I have had people ask me to psychically 'spy' and the answer is always no. You can not only get yourself into a load of trouble but also easily cause irreparable harm to others, especially if you share personal information that is not yours to share.

It was sometime in the early 1970s that I started classes on meditation and psychic awareness. After some time had passed, my instructor gave me an envelope with a picture inside and asked me to drop into a state of meditation and tell her if I saw or felt anything. Almost immediately, I was mentally watching a movie of a young woman. I began to describe her, even got her name and occupation, a librarian. It turned out I was 100% correct. We were both amazed to say the least.

What followed were years of working with clients, during which I learned and gained a deeper understanding of how it all worked. What I did with the picture was called psychometry. This is the procedure of holding a photo or a piece of metal, jewelry, or keys that belong to the person while dropping down into meditation. At this point, the 'seeker' of information from these items would begin to see mental pictures or experience feelings relating to the item held.

What I now know and understand is that the holder of the item(s) received a mental transfer of information or a morphic field that represents who they are. It can easily be read by someone who knows how to reach a meditation space that allows them to 'see' into that field. Pictures and metals hold the morphic energy fields of the people in the picture or jewelry worn by them.

I later learned that I could have a name spoken to me and receive information, as everyone's name is specific to them, their essence, which tells anyone who they are. Even more so, as time went on, my client only had to think of the person, and I could pick up on the information. By doing so, she activated the morphic field of the person simply by thinking of them.

I was entranced and enchanted with this amazing ability. As time went on, I began to get clients who had genuine problems, and I realized this was no longer just a lark. Some clients needed real work or help. Sometimes I referred them to a more qualified practitioner who could offer what they needed.

Along with all of this, I began to see and feel my own dysfunction and issues that had deeply plagued my own life. As I look back on all of this, I see that the energy of my identity and what had happened to me resonated, subconsciously, with others who came to me as kindred souls. Now that I know that, I use the Trauma Resolution procedure instead of just giving readings.

Fortunately, I was able to find a few therapists and attend some seminars that enabled me to get some respite from my past. Dr. Michael Shea's BioDynamic Craniosacral Therapy Training was a three-year residential course, totaling 750 hours, and included specific training on trauma therapy. I gained a deeper understanding of myself as I studied the work. I will also continue to work on various aspects of myself, even as time goes on. It is all good, and the journey is light-filled and not dark and scary. Once I learned how to ground myself and remain calm, regardless of the circumstances, I was able to stay present in my body and integrate whatever showed up.

Because I arranged the classes in Arizona, I had many more hours of training as we had repeat classes from which to cull our three-year group of therapists. I attended all of them and was subsequently privileged to assist Dr. Shea as well, once I moved back to Florida.

Returning to psychic states, I will elaborate on those that have become familiar to me and explain each one.

Returning to the 70s again, my neighbor and I decided to try telepathy. We obtained a book on the subject and followed the procedure each evening, where the person contacted would call the sender the next morning to report anything that might have happened.

One morning, my friend called me and excitedly told me that she thought I had appeared in her room. She then described what I had been wearing when I did the telepathic send. It was exactly what I had on that very night. Not only did she receive my message, but I was also able to send a part of my etheric body to her presence for her to see me. (You can Google the Monroe Institute for more about sending your consciousness (Bilocation) to other locations)

So far, we have Telepathy and out-of-body bilocation. Please understand that these experiences are real things. We are currently in a place and time where the veils between dimensions are thinning, so these things have become more easily achieved.

You can tune into a website that explains some extreme psychic experiences by Andrew D. Basiago. See www.ProjectPegasus.net. As a child of a CIA operative, he was privy to some very unusual experiences. YouTube will also have some excellent lectures from him as an adult.

Bob's Dead Friend

I like to include a brief story with each, as I think the explanation is easier to understand. It was around 1985, and I was working part-time in an office. One of the workers had seen my segment on TV with Ruth Rogers and asked me if he could give me a name to see what I could 'get'. Of course, I agreed. When I spoke the name that he said, I found myself in a black nothingness, floating, and I could tell that this person was in a tremendous amount of pain. When I asked this coworker, Bob, about the name, he said the man took his own life sometime in the 50s because he had such a painful disease he couldn't stand it anymore. He used carbon monoxide poisoning from a car exhaust to complete his suicide.

What I later learned is that your thoughts can specifically create an objective reality once you have moved into other dimensions. So if your thought is pain, guess what? Consciously changing your thoughts can help alleviate pain.

I told Bob that I would work on it more when I got home so that I wouldn't be interrupted. Once home, I lay down and went into a state of meditation. I realized I was connecting with someone who had died (a first for me) and wondered how I could help. (There was the intention sent out to him that he picked up somehow and made himself available for help) As I tuned in to him, I tried to send messages, but there was no response. Since I had already realized that Thoughts are Things, especially in the etheric spaces and other dimensions, I decided to send pictures. I sent a picture of a park and then another, and I could feel I had his attention. I then asked for angelic help for him to move on, and BAM, he was gone in a millisecond, and at that point, I had to take a big breath in and began to sob with joy.

I was in momentary ecstasy wherever he went. And wherever it was, it was breathtaking, literally. Of course, his pain was gone once his thoughts and awareness were changed.

This was my first Medium connection, and many followed thereafter. You have figured out that a Medium is someone who can bridge the gap between living and the dead and see or speak to those who have passed. What I learned later is that this 'place' they go to is only a different dimension. Some cultures call these people Shaman.

You may have heard the theory going around that "we create our own reality". It can't be truer than when you pass over to the other side, as Thoughts immediately show up as Things or a seeming reality in this new environment. The more conscious one has been in physical life, the more they can navigate in the new dimension of their following life form.

My sense of all this is that there is a core soul that incarnates as physical (or non-physical, depending on the nature of its incarnation) into its next life. The core self, or soul, encompasses all these personalities, and ultimately, they all reunite. This may be why some people remember past lives so clearly or connect on a deep level with new friends, feeling as if they have known them forever.

Next to examine is Clairvoyance, Clairsentience, and Clairaudience. The first, clairvoyance, is the ability to see and experience other dimensions, including past, present, and future. This psychic is sometimes literally referred to as a Seer and can indeed See on many levels. Mental movies appear simply by 'dropping down' and saying the name of the intended person. It can be seen in the mind's eye or imagination. For me, it's like watching a movie. For those who don't necessarily see, they still 'get' what information they need to understand who they are tuning into.

If you have difficulty mentally visualizing, you can practice by simply closing your eyes and recalling what your kitchen looks like, or your car, and so on. The more you exercise this mental muscle, the more it will reveal to you the psychic impressions you are looking for.

Clairsentience is like being an Empath. You can 'feel' others, tune into them. You can also have gut feelings about something or someone in your own life. Learning to trust those gut feelings can be a perfect measure of what to do or with whom. When you learn to connect to your body as a felt sense, you will gain access to a wealth of information that helps you in life.

Clairaudience is, of course, the Ability to Hear statements or messages that can be helpful in your life or your client's. I recall working with a client who was beyond my capabilities and well outside my comfort zone. It could have become dangerous for me. I could handle it. (Watch out for that one! Think you can handle it, huh? If you have that thought, it means you cannot handle it, and please refer out!) Anyway, the last time I saw him as a client, he had just walked out of my office, and I heard the loudest shout in my head, "Never see that client again!!!" Boy, did I pay attention to that one! I left him a message with another therapist's name and number. He called later and asked why, and I explained what had happened in my head. There was no comment from him after that.

Helpful Voices

On another occasion, I was becoming very ill to the point of exhaustion. I kept hearing quietly, though, that there was mold in my house. Finally, one day, I hear that same loud grouping of words yelling into my head, "There is mold in your house, and it is killing you!!!"

It turned out there was mold in my bedroom under the carpet. (Welcome to Florida) We had that taken care of, and my energy returned. I do not know who my excellent caretakers and benefactors are. Still, I thank them all the time for the help I receive during sessions and at other times when I need help or information. Clairaudience can be a real gift!

When I first learned I was a Highly Sensitive Person, both a boon and a bane (see the book by the same name, HSP), it was through that discovery that I could pick up in my body the pain and emotions of other people. This was early on in my psychic career, and one day I had two people come to see me, both of whom were suicidal. Fortunately, they were already set up with a therapist. At any rate, when they left, I was trashed. I had pulled all their inside awfulness into me. It was a bad day for me.

I contacted a fellow psychic, one who had worked for the police in finding bodies and murderers, and she told me I had been keeping their 'stuff'. I could either hug them goodbye and return it or mentally send it back. It was neither my creation nor mine to deal with. I now happily send it back. I have also taught my body to let me know when it's not mine. When I get that, I breathe it out and look to see where it might be coming from.

Before I began to set good mental boundaries, I would be trashed on the floor or in my bed a short time before the phone rang for a client to have me help them calm down. That doesn't happen now, only when I am with a client, I feel what they feel and SEE IT AS INFORMATION only, and thank my body for that and let it go. When I can feel exactly what they think, I know better where to begin to calm them down and teach them to do it themselves in the future.

As so beautifully put in Jack Carson's Gremlin books, know where you end (at the skin) and where everyone and everything else is, outside of you. It is beneficial to practice being centered in your own skin, so that when something comes at you, you are already resourced to be grounded.

Entrainment is another great tool and gift. If you get into hypnosis (see Reframing by Bandler and Grinder, who developed NLP, or NeuroLinguistic Programming, created from the work of genius Milton Erickson), you may be taught entrainment, which involves placing your body in the same position as your client's. Additionally, during any sound therapy session, entrainment is typically a given. We connect with our clients to enable us to help them, understand them, and hold compassion for their situation. Suppose you are sitting across from a client and mirror their body position, slowly and subtly. In that case, you will be on the same wavelength in a matter of minutes. See also Eldon Taylor and his numerous books and CDs on hypnosis, mind control, and programming.

Of course, the most crucial aspect of entrainment is that your nervous system is calm, so bonding with a client in this way will send them to the same peaceful place that you are in. Remember that when you mimic their body position, you will initially feel their angst, so it is essential to get and stay grounded.

I spoke earlier of psychometry. My first experience was holding a picture, but jewelry works as well as saying the name. Please understand that we are all one, everywhere is right here now, and there is no time or space. Suppose I drop down into meditation and I want to visit a time in the past or future, or a building in Nova Scotia. In that case, all I need to do is say the time I want to go to or the address of a building or a place in a different country or even hold something from a long time ago. Once your meditation skills are proficient, these things will be easy. You will get or feel the picture of where you sent your mind. Metals and other objects used by the individual will hold a morphic field of information, available to anyone who can tune into the field.

What about getting some help if you find yourself in a situation where you get stuck? Just call an Angel. No, I am not kidding. They love helping. I use Michael Archangel for removing entities, moving stuck souls forward to a higher plane, and helping with healing and other purposes. There are no limits. Over the years, I have accumulated a group of wonderful helpers. I had a psychic tell me that my field of energy was full of helpers, as she could see them when I was in the process of working with someone.

I mainly used a few legions of angels when I was asked to clean some property in Skunk Creek, Arizona. That property, I found out later, was quite polluted, and even the water in the creek was so toxic that it burned the skin if one put their hand in it. When an area is toxic or destroyed in some way, it leaves an invitation open for dark entities to inhabit it. I don't want to visit the Tower of London, no thanks. That is way over my pay grade.

Toxic Influence

As it turned out, a coworker, Joyce, told me that her little girl was so afraid in their new rental home. She couldn't abide the windows being open, and she said things were there at night, walking around. She was terrified. The mom, of course, didn't feel a thing. Please believe your children; they feel much more deeply than adults usually do. I got the address of their new place and dropped into meditation that evening. I saw horrible creatures coming from under the home, into the roof, and swirling all around the walls and windows. I brought in a few angels, but they were soon blown back away. Finally, I brought in legions of angels who were in every direction, under and over the house, and all around it. There was a silent agreement that they would be there to protect the home.

The next morning, Joyce came to the office to report that her little girl had said she could sleep with the windows open without any problem and was always comfortable in her new home. The mom said she noticed it 'felt' better as well.

Anita's Release

Mentally cleaning homes and buildings from a distance is awesome. All you need is an address and to drop down into meditation, which is simply a slower brain wave, known as alpha. Two friends of mine called me to ask for help as they were hearing walls being banged on in their connecting apartments. They felt it was a friend who had just died, Anita. Background on Anita was intense. She had abandoned her baby to be with a man who was involved with drugs and the mafia. She was addicted to some heavy drugs herself. When the police found her, it appeared it was a suicide, but when we really looked at everything, it was apparent she had been murdered.

I went to the apartment complex at my friend's request, and we all dropped into meditation. It turned out that it was indeed their friend Anita who was so remorseful at having lost her baby and her life to drugs, and this man was the one who murdered her. She wanted to have someone to witness her grief and remorse, which is what we three did. As an empath, I began to cry, more like keening or wailing. My face got very red, so they told me, and when it was over, we could feel her being settled and gone. I went into the bathroom for a few minutes to wash my face, and when I came out, my face had no redness and looked normal. I didn't even have swollen eyelids. To top it off, of course, there was no more wall banging in their adjoining apartments.

Cleaning House

A client of mine practiced Wicca, which is white witchcraft. She used herbs and incantations to help create what she wanted. I assume she had not set forth an intention to bring in only light workers as her place began to get weird. Since I lived close by, I was also able to visit her home and found that it had some dark energies there. I explained a simple way to dispel them: tell them to leave, don't ask, demand, saying they are not welcome. At the end of that, we imagined a statement in runic letters (or English, whatever you like) around all windows, doors, air vents, chimneys, and any other open portal we could identify. We imagined it to say, "All Who Enter Here Be Only of the Light". For her she needs to reinforce this once a year, especially if she moves into a different house.

Angels & Entities

Clearing entities from a person is an interesting concept. First and foremost, never do it if someone else asks you to clean another person's entities. The reason is that if they don't know, the entities can come back thousandfold. That did happen to my friend's son, who had been a meth addict. Another caveat to pay attention to is if they have indeed been on drugs, even prescription drugs, or have been in a coma or seriously depressed lately; all those events can allow entities to come in. I had a man call me to rid him of an entity, and on the phone interview, it came out that he had a girlfriend that he was stalking. I refused to work on him because with that kind of dark energy, the entities would have loved to live on in him. Obviously, stalking is not exactly a nice thing to do.

I always use Michael Archangel to assist me in this, as well as to help someone who has passed and is stuck to move on. A woman came to me for a session and primarily wanted to get this 'icky' feeling out of her. When I 'looked' into her, I could see whatever it was swirling through all her body. I called in Michael and then began my own work. As I began to pull it out of her, I could see where it was coming from in her body, but I didn't mention it to her. However, during the entire process, she could feel it coming from her arm, chest, or hip, wherever we were pulling it from. At the finish, Michael took it away, and we filled her with light energy.

Past lives can affect the present life. I have found only a few clients in these situations, but I don't discount this possibility. I had a massage client with a sore leg. I worked and worked on her to no avail. During one treatment, I thought I saw in my mind's eye a large knife, like a machete, in the very part of her leg that was bothering her. That day that I pulled it out, she never experienced pain there again.

Tears in the energy field

Another book I recommend is The Holographic Universe by Talbot. In these two examples, I will speak about holes in the morphic field or holographic field of the body. I was giving a massage one day to a new client, and when my hands were on her back in the kidney area, I had the strongest urge to leave them there. While I was doing that, I thought she might think this is weird, but if I even thought about lifting my hands, it felt so wrong in my gut, and the imaginary sound of chalk screeching across a board came to mind.

We finished the massage, and neither of us mentioned the extended hand position. I heard from her a month later, and she said that she thought whatever I did to her back gave her energy back. (She had been diagnosed with chronic fatigue syndrome). What I realized is that her adrenal area, above the kidneys, must have had holes in her energy field and she was leaking out all her energy.

A second example is a young woman who was sent to me by her mom. She was depressed and had no energy. As I was doing her reading, I came upon a mental scene where she was a young girl, and someone was speaking to her in a way that was uncomfortable for her at some public gathering. That told me it was about then that her problem began. I saw a picture of her closet at home, all black, and that she owned a belt of every kind. I asked her if she specifically chose to wear black every day, and she looked blank for a moment before agreeing, saying, "I guess so, yes." I noticed she was also wearing a belt and asked again if she wore belts all the time. A puzzled look appeared on her face, but she agreed again when she thought about it and mentioned that even her robe had a belt.

As you can see, we can easily have rents or tears in our energy fields. We will leak energy without knowing why we feel so tired all the time. Having these unprotected spaces around our fields of energy, we then unconsciously allow other energies to seep into our bodies, many times causing mood swings, depression, and so on. Once you learn how to scan your own body, you can become aware of these trouble spots and address them.

Now, back to our client, this is a good example of SC work where the body holds the Trauma. I asked her how it would feel if she were wearing a yellow dress with no belt. She immediately went into panic mode, which was quite obviously evident on her face. To calm her down, I mentally brought her back to her safe zone, which was characterized by black clothes with belts. What I then explained to her was that whatever had happened had torn a hole in her energy field.

Together, we worked to seal that waistline tear, with me doing it energetically and her picturing herself as whole and sealed up. By the time she was ready to leave after our session, she could think of wearing a blue dress without a belt without feeling panicked. We had literally sealed the problem, and from then on, I felt confident that she would continue to improve.

Another friend of mine had taken the Jose Silva Mind Control Class. In a few minutes, she taught me how to body scan. Again, get to alpha brain waves. Picture the outline of a body and mentally say the name of the person you want to monitor. Place their field onto the form and slowly work your way down the form with your mind. You will find that you 'run into' places that encourage you to stay or pay attention. When I see that, I start working on the area. This seminar is quite good, and I recommend it to anyone interested in exploring more esoteric work. His book, Silva Mind Control, is an excellent read to prepare for his seminar.

Manifested Anguish

What follows is a poignant experience I had some years ago. I had a relationship with a man who had some humiliation and demotion at his job. I could see one morning that he was particularly suffering. I said to him that I could see this and said that if he needed my help in any way, I would be there for him. The day went on, and soon we were both home from work. I was cooking dinner and wondered where he was and went looking for him. He was lying down in our bed with the windows open and the overhead fan on. It was a lovely spring day, about 6 P.M. or so. His eyes were closed, and as I looked up at the fan, I could see translucent circles swirling around as the fan caught them. I could also see them emerging from the center of his body and rising to the ceiling. I could not imagine what they might be. I said to the man that I was going to lie down next to him for a while. Then the circles increased and filled the whole ceiling in the room so much so that some came down and began bumping into my face. I could feel this energetic manifestation and watch it as it was affected by the moving fan. There was so much anguish in him that it was physically palpable.

I knew I had to get up and get dinner on the table, or he would become suspicious, so reluctantly I left the room. We soon sat down to dinner, and I felt strange, not wanting to eat. I told him I was going to go back and lie down. He then finished eating and went out to cut the grass. The circles in the room were gone, but I began to empathize with all that energy that had come out of his body.

Meanwhile, I was experiencing elevated activation, where I started to cry, all the time knowing that this was an empathic experience I was helping to clear. Thankfully, I didn't own this and learned to split my attention, knowing I was OK yet allowing the energies to pass through my body. I began wailing almost into a scream. I was aware, yet at the same time, stunned that this sound was coming from me. The man came back into the house and ran into our room, asking me what was wrong. By this time, it had been long enough and so intense that I could barely speak.

I was able to get out the words, "This is yours," and began to calm my keening. I saw him hit the wall behind him with his body and slide down to the floor. He crawled out of the room and left. It took me some time to get grounded. I called a friend to connect with so that I could help him and then find him. When I was finally able, even though I felt like I had had a 220 line run through my body, I left the room and found him on the floor of the bathroom with his face in his hands and crying, sobbing. I said a prayer and left him to his own devices.

The next day, we were in the car together, and he said, "Thank you for what you did for me." I left it at that as it seemed personal and deeply traumatic. He also improved significantly as the days passed.

Here I am pointing out some pertinent things. That morning, I set forth an intention that his body heard, offering any help he might need to get through this difficulty. On some deep level beyond conscious awareness, his body knew that he could get relief with that offer. His body had packed in so much energy from this experience that it came out visually and was affected by a physical fan, which I felt as I lay there in a room filled with his releasing energies. It was one of the most amazing empathic experiences I have had to date, even these many years later.

When he came back into the room, he saw my emotions and recognized them as the ones he had been holding inside himself. I don't mean he was able to say, "Oh, that is what is going on inside of me", no, but his body recognized the energy waves as being the same as what he had held. My body became the link to the emotion that had passed through me from him, and as he saw it, was able to own it as his, and then finished processing the grief and emotions. The circuit was complete and resolved when he finished releasing his final tears. The Trauma was integrated.

.

From Dragons To Butterflies –
Trauma Resolution & Morphic Field Energy Healing

Chapter 7

TRAUMA RESOLUTION

What exactly is "Trauma"? According to Dr. Michael Shea (2007), "Trauma is a physiological or psychic response to physical, emotional, or spiritual injury. It is usually the result of being overwhelmed by shock and not having resources available to integrate the shock; thus, Trauma occurs and becomes embedded in the mind-body continuum."

As you may have read in other chapters of this book, Trauma can be re-remembered in the body by something as simple as a smell or a touch. Suddenly, the individual may feel faint or nauseous and need to sit down. There can be other body sensations too numerous to mention here, but the idea is clear. After this passes, they go on with their day or life, never knowing that the body gave them a momentary opportunity to integrate whatever Trauma was triggered by a seemingly innocuous reminder that brought on bodily sensations.

The Body Holds the Trauma

This example may help. While in my mid-20s, I found a book I couldn't put down, Watership Down by Richard Adams. The story is about a warren of rabbits living in a forest where a construction company will soon be destroying the land to build a large structure.

Of course, the rabbits didn't know what was happening with the loud machinery and the collapse of some of their warrens. Mr. Adams assigned all the rabbit character names, and some even possessed special sensory abilities, as well as precognition.

About 20 years later, the animated version of Watership Down was on TV, and my husband and I decided to watch it. At the end of the movie, I got ready for bed, and my body could only go into a fetal position and sob and cry. What is important here is that my husband knew about my childhood trauma and knew that any abreaction I had was not to be interrupted by him, just that I had asked if he could be an observer and hold space for me to finish. He indeed followed all my instructions (that had been previously set up) so that I was able to integrate whatever Trauma the film brought up in my body.

YOU CANNOT GET THROUGH TRAUMA WITHOUT GOING THROUGH THE BODY!

This is the most important thing you will learn in this book. Hopefully, you will learn how to integrate Trauma by reading these chapters.

While I was abreacting (being in a complete emotional and feeling state of sobbing and distress), I was clearly conscious, thinking that this was really a significant experience and that I knew it would be over soon. Yet, it also felt OK to be crying so intensely. (Dual attention) Mentally, I saw myself floating in a dire, dark space, all alone with nothing around me. The feeling I had was that soon I would be snuffed out forever, as if I had never existed at all (annihilation). At the same time, I knew it was not true and was aware I was safe.

When it was all over, my husband and I just wondered what had happened, but we went on with our days. Some years later, I was in a class with Dr. Shea, and he was talking about trauma reactions. I asked him about that movie and why an animated film would evoke such a strong reaction. He said the theme of the book and movie was about annihilation.

At that moment, I realized that when I was being abused as a child, I felt as if I was going to be disappeared into nothingness, literally annihilated. Because I was so young when it happened, the bodily feelings were implicit, meaning the memory has no language and is only held in the body, with no words to explain the sensations and emotions. This revelation became a powerful piece of the many puzzles that had been plaguing me my entire life.

Fortunately, I was able to stay present and observe my abreaction with dual attention, allowing myself to cry and knowing that this would soon be over. I was not being hurt at this time, just clearing the body from a body memory. I knew I was safe, which allowed me to process safely. I could sense that these sensations were from the past, and in that present moment, it was not happening, just a memory.

You should see that I was also able to resource myself. I had previously arranged for my husband to be present, staying calm, and not touching me or trying to calm me down. I had a witness there to help me hold the Trauma and see me through it. As an afterthought, not every partner can have the presence to do this for you. I am fortunate that my husband can. Remember not to be critical if your partner says that they won't be able to hold space for you during trauma memories. Consider seeking out a therapist or a trusted friend for help instead. I believe that because my husband also works with Trauma in his HK practice, he can be present for me.

Resourcing is significant. This can mean anything from feeding your soul with a good book or a warm candlelight bath to singing in a choir. Nature is the true soul healer, so parks, beaches, and your own backyard are usually readily available. Finding a friend who will not try to fix it for you but listen with their heart space open to help hold whatever you might be going through. Make sure you eat good food, stay hydrated (drink water), get enough sleep, and exercise to the best of your ability. If you need a therapist, inquire around to find one that is just right for you. Being present, listening, and asking thought-provoking questions also allows you to become aware of what is going on for you. Massage therapy, Yoga, dance, Zumba, and exercise of any kind are also necessary, as moving the body can help move out blockages.

It is also essential to realize your inner resources. Are you resilient, loving, intelligent, capable, emotionally stable, and courageous? Or do you have a skill set that is uniquely yours, developed to a high level of competence? It may take you a while to tap into all your inner resources... don't be afraid to ask friends and loved ones what they think your inner reserves and gifts are.

When we prioritize taking care of others before ourselves, we become even more depleted. Our cup is empty, so we can't even support ourselves, let alone someone else. How can you fill your 'cup'? Take charge of your life and resource yourself. Surround yourself with beauty; having a pet is wonderful. Get support from friends, therapists, or even your spouse or partner. Get out in nature and experience the peace and calm of the growing things in our world or bodies of water. Bodies of moving water, like rivers and oceans, can add negative ions to your energy field (good ones), which dispel the positive ions of stress and emotional blockages. Good, nourishing food is also essential.

I have also used EFT, also known as Emotional Freedom Technique. It is a tapping procedure to allow the body to calm down. You can Google EFT and even download a free booklet about it. There are also trained therapists to help you learn how to use this. Read below how I have used it for panic attacks.

Panic Attacks

I used to wake up with panic attacks. If you have ever had them, there is a feeling that you are sure you are going to die, with your heart beating so fast, as well as other awful symptoms. I knew it was not true, and I was having a physical reaction, so when it happened, I used the tapping method (EFT) to calm my body down and reduce the panic. It may take an hour or more, so be patient. Just keep tapping. Instead of following meridians, I just tapped on my sternum or my leg. After so long, your EFT hand may get tired, so trade off on spots to tap. This grounded me and allowed my SC mind to know I was safe, and the symptoms subsided.

Rue Ann Hass, M.A. (see IntuitiveMentoring.com), and her four books are wonderful tapping aids for those of us who are highly sensitive to everything. She is also listed in the book reference section. Her website lists all her books.

It is important to notice other body difficulties, for instance, chronic constipation, body pains, autoimmune diseases, insomnia, depression, chronic headaches or pain, experiencing continuing car accidents or falls are also important indications of Trauma and stress held in the body. Addictions and compulsions are something to make note of, as well as one destructive relationship after another, and a pattern of needing constant upset and disharmony in your life.

Examples:

I worked with a client who complained of a sore ankle that had plagued her for years. Initially, she had had a motorcycle accident where she was hit and thrown up into the air and landed in the middle of the road. We did a trauma session and found that the ankle held a great deal of the accident memory and injury. Once we cleared and integrated the Trauma, she later told me that her ankle pain had finally resolved.

A friend of mine had two knee injuries within a year and was again on crutches for 6 months. She is young and healthy, so the knee not healing was a good indication that it was holding something else. She met with a healer locally who mentally saw a little girl holding her knee and crying. The healer said that she felt the woman was still carrying some sadness and Trauma, and it was in her knee. It was also true that she had had a traumatic childhood. Once she realized the patterns, she was empowered to make changes.

When I worked with my friend, I showed her how to ground and center. That is the most important, so that whenever emotions and pain showed up, she could get herself into her body and settle. Her husband also assisted by getting her a book by Dr. John Sarno about how we hold body pain from Trauma and how to release it. She viewed YouTube videos with Dr. Sarno explaining the concept of pain, and by the time she finished them that day, her knee was pain-free.

Then it would show up again, so she called to see if we could resolve the remaining issue. When I worked with her to finish up resourcing herself, we set up mental scenarios where she could imagine being in a place that had wonderful memories and bringing in her husband to support her (as he does also in real life), and finally, Dr. Sarno.

They all held her hand as she felt the pain, and soon it was gone. If her pain came back, she had a resource to ground herself and connect to two important healing people in her life as supports. Within a short time, she was able to put the crutches away for good and is now able to walk pain-free. Unfortunately, her childhood family is very dysfunctional, so she has little contact with them, which turns out to be healthy for her. She is now happy with this arrangement and has two wonderful children and a fantastic husband who truly loves and supports her.

Suppose you are a therapist working with a client. In that case, it is your responsibility to notice any bodily activations, such as shallow breathing, sweating, flushing of the face or neck, hand wringing, sudden leg or arm movements, and so on, during a session. It is essential to bring these to the client's attention and keep them in the present moment, reassuring them that their body is reliving an experience and that they are currently safe, sitting with you. Have them describe what they are sensing. Make sure to stay away from emotional triggers, such as fear, anger, etc., but focus on body sensations, like sweaty hands, difficulty breathing, nausea, and so on.

Fortunately, when conducting phone sessions, it helps to be an empath and to sense the feelings of my client that something is going on. It is almost a sense of heaviness, as if something is in suspended animation, and if I also empathize with bodily sensations, that provides even more information. The reasoning here is that the phone client may sometimes become disconnected. If I can be made aware of something 'going on' with them, I will ask, "Are you OK?" which can bring them back so they can explain what they are sensing.

I will mention this again in the step-by-step trauma resolution chapter. You will be working with the SC mind as the client experiences these body sensations; if they remember the experience, all the better. Suppose it is only implicit (pre-verbal). In that case, they can call it up by mentally or verbally saying, "Bring back the memory." To clear all the Trauma, it is essential to allow the client to settle down after experiencing the body sensations for the first time. It is best if the body sensations can be allowed to stop, resolve, and the client can be at peace. It is rare for the first activation to be totally resolved. Therefore, you bring the client to a peaceful place and then return to the Trauma later to test how the body sensations now respond to the memory. (The best book for this is Babette Rothchild's The Body Remembers) Trust that with each return to the memory, the body will have less and less response. Ideally, you want no bodily sensations, and then you will know that the memory no longer activates the client. It simply becomes "something happened" in their mind, and the SC mind knows the person is safe and free from the Trauma.

It is essential to keep them in their body and not allow dissociation to occur. If you focus on their face, you will notice dissociation. Sometimes I tell them that I may put my foot on top of their foot to ground them (make sure you have permission to touch them on their foot first), or I will ask them to look at me and focus on my eyes. Since I am grounded, it will help to ground them, and all the while, they are going through body sensations.

If they begin to abreact, break away from the subject at hand and talk about something light until they are settled. Then let them know you both will be returning to the issue to clear and integrate it into their body. I usually like to let them know at the beginning of the session that we might take small breaks to allow them to settle. Telling your client how the session might unfold is a way to reassure them that they will be safe and you will be there for them. An excellent reference for this work is the book "The Body Remembers" by Babette Rothchild (as mentioned above). For practitioner training and to search for local therapists, see www.bodynamics.com.

Peter Levine (Waking the Tiger and Healing Trauma) is the developer of Somatic Experiencing ©. Please visit www.traumahealing.com to view his teaching schedule as well as a directory of practitioners trained by the Foundation for Human Enrichment. His book, Healing Trauma, comes complete with a CD and instructions for body-centered healing. This is a small yet profoundly powerful book for those seeking to embark on their own healing journey.

If you are a craniosacral therapist and prefer to use this model, please get in touch with www.michaelsheateaching.com to arrange training in the BioDynamic methods of relieving Trauma in the body. Dr. Shea has written 4, soon to be 5, volumes on Biodynamic Craniosacral Therapy that contain valuable information on all levels of the body, mind, and spirit.

If you choose to read Trauma and Recovery by Judith Herman, be warned that the stories included are intense, especially rape, incest, and PTSD associated with war and witnessing horrific acts. As a matter of fact, Dr. Shea himself was in the service. He was stationed where a terrorist had set off an IED, which resulted in a fatality that was a terrifying trauma for all involved. Following that incident, after he left the service, he found that he was unable to hold down a job. He exhibited some behaviors that were less typical of him. Fortunately, he was able to get therapy and resolve that Trauma.

Dreams can be an essential indication of a hidden issue to be explored. It was Sigmund Freud who first recognized that dreams can be an indication of the SC mind trying to tell us something. I have also mentioned The Inner World of Trauma, Archetypal Defenses of the Personal Spirit by Donald Kalsched to understand how dreams and fantasy images can also be important in the healing of Trauma.

Example:

Some years ago, I had a horrific and vivid dream where I was standing knee deep in water, and body parts were falling from the sky. There was bloodied water around my legs, and I could even feel the body parts hitting my legs. I was horrified. I set up a hypnosis session to uncover the root cause of this dream.

When I was two years old, my parents divorced. They had had a very violent two years in our house, which involved throwing dishes, fighting, violence, and loud, yelling arguments. The hypnosis showed a scene where I was in another room, and my parents were in a bedroom arguing about who was going to get custody of me. The image of them tearing me apart into virtual pieces was clear, and I immediately knew this was where the nightmare had come from. Once I knew that was the core of the problem, the nightmare never returned.

Triggers

Anything can trigger a memory, causing the client to start showing signs of dissociation and Trauma. A word in a sentence can be the key to sending them into this state. Being grounded and present at all times will help you become aware of this state and adjust your tactics to support integration.

During a group class, we had a psychotherapist begin to tell a story of a mugging she had experienced in another state. I also had worked with one of the participants there who had been raped at machine gun point in another country when she was a young teen, and she had never told anyone about the incident. At the time of discovery, I provided her with the name of a reputable therapist. Still, she had only gone once, so when the instructor began to tell her story, I could see my classmate dissociate and barely be able to take a breath. I brought this to the attention of the TA, and she referred the issue to the other instructor for resolution.

I later spoke to her and ensured she understood that she could not keep this 'secret', and that her husband and parents needed to know what had happened to her. We worked on the best way to tell them, prefacing her reveal with the understanding that she was OK now and they needed to know her experience. Holding the intensity of such events does not heal them; instead, it continues the difficulty within the body. Relating such events may be necessary to have the therapist there as well to support the client during the telling.

Suppose you are an empath and feel their body sensations simultaneously. In that case, this can also alert you to pay attention to their body cues of dissociation and activation. Remember to thank your body for the information, as it is simply a means of conveying data. You DO NOT want to own what you are empathizing!

During one of my ongoing massage classes (which I have been teaching for three years), one of the participants was unable to be treated on the table. We set up a session to work on whatever it was and discovered her body had a memory of an abortion when she was a young teen. Lying on the massage table reminded her of lying on the abortion table. We worked to release and integrate the body trauma. Still, following that, it became necessary for her to feel safe being back on the table slowly. Initially, she sat nearby, observing a session and staying grounded and engaged. Soon after, she decided to lie underneath the table for a few sessions and finally was able to lie on the table without any replay of Trauma.

It is essential not to allow your client (or yourself) to dissociate. If their eyes close during an intense crying session, it is doubtful they have split attention, knowing they are safe in that moment. They are going back and fully reliving the Trauma. That will recapitulate the Trauma and drive it deeper. You may have to call them back by saying their name out loud and asking them to open their eyes. Then it is up to you to show them how to settle and calm down.

We have briefly gone over implicit body trauma memory. That is the experience of Trauma before there is speech, which is why it is called implicit. It is inside and cannot be named. There are no words, only sensation. Sometimes there is a mental picture of the Trauma, and sometimes there isn't. I was being Rolfed, and when the therapist began to work on my pectoralis major (upper breast muscle), I immediately began to sob. The picture I got was of being held down during a trauma. I breathed through it, released, and integrated the experience.

You will notice that I repeat themes and similar examples throughout. Repetition is an effective way to learn, as I have discovered over the years. The groups I teach this work to come back time and again to really 'get' what to do. This is not necessarily a step-by-step process, although it can be presented in that manner. Sometimes it is flying by the seat of your pants, and all the resources you may have at hand will be necessary to help yourself or your client resolve the issue.

Overwhelm

Suppose you already feel overwhelmed before starting your day. In that case, it is likely a good indication that your psyche is balancing many balls of energetic interference that you are trying to manage, even if doing so is not a conscious effort. The indications for overwhelm can be not wanting to be in crowds, loud music, or other sounds can be 'too much' to handle; you don't like an extensive list of to-dos, as you see it all at once, rather than one item at a time. This can also define a highly sensitive person, and the two might go hand in hand. If you are overwhelmed a great deal of the time, it is pretty assured you are holding Trauma in your body.

As I mentioned, my husband is very calm and peaceful. He works from home for a company that delivers large sets of blueprints for him to work with. Sometimes he has 10 or 20 prints lined up with deadlines. I can look at all of that and feel the overwhelming necessity of 'getting it all done now'. Thank goodness I am aware of not taking that feeling on as my own. Seeing him be calm and peaceful about the work, taking each print on with ease and patience, allows me to have the example that overwhelm does not have to be a general state of mind. Usually, an enormous sigh of relief is my final experience when thinking of the work he does. Rome was not built in a day; as it's said, only one brick at a time. Sigh, exhale, and smile.

Gating Mechanisms

A lack of gating mechanisms often accompanies Overwhelm. That means EVERYTHING that happens in the world of that person goes directly into their body, limbic system, and heart space. They don't know how to observe but take to heart all that they hear or see from starving children in other countries to a neighbor's long-lost relative's passing that they never met. After a short time of experiencing the entire world, all the news on TV, and their own lives, they are worn out and want to retreat into a safe, dark cave and never come out.

Getting embodied in your own body, feeling just yourself from the center outward to your skin, and observing the outside world as just the outside world, without allowing it to intrude and overwhelm you, is the key. Suppose you spend most of your day dissociated. In that case, the empty shell of who you are becomes a receptacle for whatever that empty shell allows in. Dissociation may constantly help you avoid facing the Trauma held in your body. Still, it also prevents you from experiencing a life you could be living.

Incest, Rape, & Sexual Abuse

There is a misnomer regarding incest. Anyone I have spoken to or who has presented with a history of sexual abuse does not realize that childhood incest does not necessarily mean that the child experienced physical coitus. Incest can be as seemingly subtle as a touch on the shoulder to the child when the touch has a sexual intention. Penetration, unfortunately, does happen, and the child's physical body is severely harmed, as well as the emotional Trauma. Incest can even occur if a perpetrator has a sexual conversation with the child (obviously not with the child), even if the child does not understand the words or intention, they will feel it in their body.

ANY form of incest or sexual molestation is Soul Wounding. In the world of the Shaman, there is a piece of the child that has broken off and is lost somewhere. In fact, during a Shamanic Ceremony, the Shaman will enter a trance and search for those parts of the child/adult, then return them to the body with a gentle breath into the Heart Center. (See Mending the Past and Healing the Future with Soul Retrieval by Alberto Villoldo, Ph.D.)

If you begin to feel activated by reading the above paragraph, please discontinue reading and seek out your therapist or someone to counsel with. If you recall the grounding and centering practices from this and other chapters of the book, practice them now. Peter Levine again delivers a great CD set, Sexual Healing: Transforming the Sacred Wound, which can be ordered through Sounds True Recordings or, of course, Amazon. This CD set is only the beginning of the healing process; however, the exercises are an excellent starting point towards containing the Trauma and achieving resolution.

Did you ever notice that looking at someone gave you a creepy feeling? I completely trust my instincts and steer clear of them. It does not matter if they are handsome or beautiful; it is the slimy feeling you get from their aura. Any sexual abuse from them can also leave that same feeling of sliminess on the victim. It will, however, usually be localized in a particular area and not absorbed into the entire person, as is typically the case with a perpetrator. This is not judging someone else; it is being aware that someone could be a potential harm.

I was in an Incest Survivors group for seven years. It was at that seven-year point that I realized we called ourselves survivors. We named ourselves survivors, and the picture I got from that was just barely slogging along in the world. We were not thriving in the world; we were just barely surviving. I believe I gained a great deal from the group and was healed in many ways; however, at some point, there comes a time to say, "Yes, I survived and now I want to move on in the world and Thrive in my life." Victor Frankl's book, Man's Search for Meaning, demonstrates that getting up and Living, no matter how dire the circumstances, will always embolden the spirit and bring Life to one's Life.

There are many other books about surviving sexual abuse. I would suggest reading the works of those who specialize in healing, especially Peter Levine. For me, I was just overly activated without knowing how to contain and settle. I didn't know how to put the book down and integrate the body sensations—learning that first is most important. My mother was incested by her stepfather. At the time, there was no understanding of how to work with sexual abuse. She was at her psychiatrist's office, on medication, and often with nervous breakdowns. She didn't know how to contain her Trauma. She acted out her sexual issues over and over as her subconscious mind tried for resolution. She was on a hamster wheel going nowhere.

It is also essential to understand that it is not necessary to remember the details of any sexual abuse. The body remembers, and when the sensations of the Trauma are worked with in a therapy session and cleared, it does not matter if the client remembers what happened. I had a client some years ago who was molested as a child and had her dolls all around her during the abuse. When we discussed her remembering only that, we agreed that her dolls held her secrets, so she wouldn't have to carry them as the child she had been. She decided to imagine a doll that held everything for her. We thanked that doll, and my client invited the doll (in her imagination) to whisper in her ear all that had happened to her. That was the key to allowing the client to embody the abuse without being retraumatized by it. Of course, there were no words, but just the imagined

162

transfer allowed the client to be OK holding the memories.

Sylvie's Session

When I was practicing as an Intuitive Counselor in Arizona, I had a client who came to me seeking insight into how her court trial would turn out. I am not gifted with precognition, so I told Sylvie that there was something more we could do.

She was being sexually harassed at work by her boss. As a footnote here, whenever someone in power goes after someone to make them a victim, the person who is being sexually abused (yes, sexual harassment is sexual abuse), there will be a weakness in their body, even to the point of collapse.

I explained to Sylvie that we could empower her to the point of strength where she could look at him in the courtroom and not be intimidated by his glare back at her. So, we began to work. I had her imagine he was there, staring at her, speaking his words to make her afraid. We went through her body sensations until they were cleared. It was a good hour before she felt settled enough that his name could be spoken; she could see him in her mind and hear his voice without becoming activated.

As the final empowering moment, I told Sylvie that I was going to put his energy in my right arm. I knew it would do me no harm, as I was aware I could easily clear it. Bringing his slimy energy into my arm was indeed a creepy feeling. It was a morphic field click-and-drag thing, if you get my meaning. I imagined his arm's field and pulled it into my right arm.

Then, before I took any action, I explained my plan. I was going to put that hand on Sylvie's left leg. I was going to be strong about it and hold tightly. Then it was up to her to pull my arm off. I was not going to make it easy.

We did this slowly so she could feel the energy of him in my arm and hand as it gripped her leg. She reacted, and we waited until she centered and grounded. Then, she grabbed my hand and arm with both of hers and pulled his (or my) hand off her leg. Her relief at having the power and strength to do that was audible and clearly a triumph for her.

Following that session, I heard from her that her lawsuit was successful in her favor, and she was able to be in the courtroom without feeling like she either needed to crawl out the door or was inactive. She only felt the power she now knew she had over him.

Transference and Countertransference

What follows is an excellent explanation of transference and countertransference. It is directly from www.clinpsy.
Org.uk/forum by Dr. Dot. The website has many more good points to examine.

My personal comment on this explanation is that if the therapist is not aware of any transference or countertransference, both therapist and patient could become seriously at risk. Please note the story following Dr. Dot to understand my concerns.

"In a therapy context, transference refers to the redirection of a client's feelings from a significant person to a therapist. Transference is often manifested as an erotic attraction towards a therapist. Still, it can be seen in many other forms, such as rage, hatred, mistrust, parentification, extreme dependence, or even placing the therapist in a god-like or guru status.

When Freud initially encountered transference in his therapy with clients, he felt it was an obstacle to treatment success. But what he learned was that the analysis of the transference was the work that needed to be done. The focus in psychodynamic psychotherapy is, in large part, on the therapist and client recognizing the transference relationship and exploring its meaning. Because the transference between patient and therapist occurs on an unconscious level, psychodynamic therapists, who are primarily concerned with a patient's unconscious material, use the transference to reveal unresolved conflicts patients have with figures from their childhoods.

"Countertransference is defined as redirection of a therapist's feelings toward a client, or more generally as a therapist's emotional entanglement with a client. A therapist's attunement to his own countertransference is nearly as critical as his understanding of the transference. Not only does this help the therapist regulate their own emotions in the therapeutic relationship, but it also provides the therapist with valuable insight into what the client is attempting to elicit from them.

For example, suppose a male therapist feels a powerful sexual attraction to a female patient. In that case, he must recognize this as countertransference and examine how the client is attempting to elicit this reaction in him. Once it has been identified, the therapist can ask the client about their feelings toward the therapist and examine the feelings the client has, as well as how they relate to unconscious motivations, desires, or fears.

There is always a level of transference and countertransference, as therapy is a relationship. As therapists, we have a vested interest in our role and want to provide a valuable intervention; we care. We have feelings for people with whom we have relationships, whether they are clients or therapists. Those feelings can be positive or negative. The very reason that has brought someone to see us may predict some of the likely transferences that may occur. Likewise, some countertransference can be predicted sometimes from our own histories."

I had a friend many years ago who had severe emotional problems. She told me that in one experience, she had gone to a psychiatrist for many years. She trusted him and appeared to be making good progress in therapy. One night, he unexpectedly came to her home. He asked to be let in and proceeded to rape her. I'm not sure if she will ever be OK from that experience, given that it was added onto everything else that had happened in her life. Unfortunately, she was a victim and again was victimized by the very person who was supposed to hold sacred space and safety for her. There was definitely a transference-countertransference problem in their relationship.

Hypnotherapy

I want to speak again for a moment about hypnotherapy. This therapy can be most healing and clearing. It is essential to find a hypnotherapist who is trained and has extensive experience in the field. Please also notice if you feel comfortable with this person.

Hypnosis and hypnotherapy have been seen as mysterious, magical, religious, and power stealing. Not true. In fact, when you are hypnotized, you can stand up and walk away. You are peaceful and comfortable, and you experience what is around you just as you did before you were hypnotized. Some therapists can give commands for you to forget what happened in the session, and that will be something you want to discuss with your therapist of choice.

Two of my favorite hypnotherapists are Milton Erickson and Dave Elman, both of whom lived many years ago. Dave Elman was not classically trained, as far as I remember, nor was he licensed. However, doctors sought him out to hypnotize patients so surgery could be performed with no pain. He was a magician if there ever was 'magic' in hypnosis. Some can also do that today. Milton Erickson was a gentle genius in his own right. He would be called to do house calls in the most extreme circumstances.

I remember reading about a case of an older woman who was extremely depressed. She lived alone. When Erickson worked with a patient, he gathered as much information as possible about them. He entered the lady's home and noticed many beautiful violet plants. He also discovered that she was an avid churchgoer and quite involved in activities and with the parishioners. As they were talking, he suggested to her that she might consider how happy it would make her to be able to share her beautiful plants whenever a birth, wedding, or other occasion came up at church. He suggested that she would feel great about sharing these beautiful plants as gifts for special occasions. Just the way Erickson would speak was hypnotic in and of itself, and his patients fell under his suggestive spells. He didn't need a special chair or a watch to wave in front of their face; he just spoke comfortably and softly after getting to know the patient.

In another instance, a patient was in great pain with cancer. There seemed to be no medicine to alleviate his pain. The man had been a vegetable farmer for years. After they had spoken for a while, Erickson began to talk about tomato plants and how the seeds start to germinate in the dark, rich soil that has been loosened for the roots to grow comfortably and cool. He described the process further until the entire plant had grown to maturity, all the while emphasizing the comfort and ease with which the plant's growth occurred. When he was finished speaking, the man's pain was gone entirely. He had empathized with the plant's comfort and ease, bringing that into his own body.

A patient came to him, saying he was sure the aliens were going to come and take him to their planet. He told him to bring food, water, and a sleeping apparatus and go to the mountain where he thought they would come to get him. The patient did that, and after a while, he came down from the hill. When he saw Dr. Erickson again, the doctor asked him what had happened with the aliens. The man just said, "I think I was a little delusional." Erickson didn't argue with him or tell him he was crazy or try to fix his delusion with any other therapy. He validated it and let the man prove it untrue for himself.

From Dr. Erickson's brilliant work, Neuro-Linguistic Programming (NLP) was developed. This is also an avenue you want to research, as it is pretty effective with clients.

Chapter 8

How to Work with Morphic Energy Fields – Instructions

Before you begin working with Morphic and Energy fields, your hands need to be sensitized so that the buzzing feeling can be automatically accessed. You practice this by first rubbing your hands together vigorously. You will then hold them an inch or two apart and slowly move away and towards the two hands. You should be able to feel the buzz at this point. Practice this regularly by sensing the energy fields around trees and plants.

Always move slowly towards the focus of your practicing item so you can feel where the field begins. If you go too quickly, you will go through the field and wonder why you cannot feel it. Practice on your own field, going down your arm about an inch or so away. If you have a friend who will allow you to run your hand down the front of their body, that will be excellent.

What you eventually want to accomplish is just a thought that will 'turn on' the energy field in your hands. My hands know when I am about to engage their energy sensing and begin, even while the client is still speaking to me, before the session starts.

The next step is to be able to extend energy fields from all your fingers. After you have become proficient with the palm of your hands, begin to point your four fingers toward a tree and sense the field. This will be very useful if you want to clear a field in a client using the tips of your fingers. Sometimes I do this before I remove the morphic field of the organ itself.

Here is an example. I might be scanning a kidney and the ureter. (Remember to have an anatomy book available for reference.)

If you have developed body scanning (Jose Silva's **Silva Mind Control** book and training) you will be able to see or at least 'get' that there is a problem in the ureter. Say I find the energy field of a kidney stone lodged in the ureter. I can be on the phone or have my client in front of me; either way is acceptable. At this point, I am holding the imaginary or visual field of the ureter and the kidney stone in front of my mind's eye.

I then point my finger or fingers at the stone and make the connection. It will be like dragging and dropping on your computer or cell phone. Once you connect to the stone, begin to slowly guide it to descend towards the bladder, always being mindful of the client's reports of any physical sensations they may be experiencing. You will want to proceed slowly and do your best to reach your destination. If it gets stuck, you may need to add a small amount of energy or make a gentle finger movement to get it moving again. It is essential that they stay grounded and genuinely experience the sensations of what you are doing. That is a given with every energy session you conduct. It is necessary to obtain client/patient feedback during every session.

On occasion, I may mentally ask, What other type of energetic therapy can help this problem? Or, what else can be helpful here? An herb (always etheric, do not diagnose or prescribe) or a rinse with energy or fluids that might facilitate the expression of the problem. This, of course, is all done with your mind.

A friend of mine used to clean out my chakras. She would stand behind me and 'see' the first chakra, then send pure water, light, or whatever the chakra indicated it needed. I would also tune in to see what it looked like and what she was doing. One time when cleaning a chakra that was particularly clogged, we both laughed out loud as I told her I thought I might need to vacuum the room after this one was cleaned.

Practice Session

You have your volunteer friend in front of you who understands what this energy work is all about. They have been coached in grounding, and they will need to let you know what they sense as you work. You will be seated across from them, with large pillows in your lap or behind a small table, allowing you to rest your hands and arms as you hold the morphic field and will enable it to balance.

You can begin by body scanning, either with your hands or using the Jose Silva method, to determine where to work. Alternatively, you can ask your client what is bothering them physically. You then want to confirm and, with the palm of your hand or extended fingers, go down the field and verify where you feel the most congestion. If it is the same area of your friend's complaint, you can begin to work. If it is a different area, you must first determine what needs to be worked on.

Remember, all cells, organs, and tissues are conscious. You will mentally or audibly ask the morphic field of the organ to come into your cupped hands. Meanwhile, you have your arms and hands resting on the table or pillows. As the organ's (or muscle or bone) field comes into your hands, you will feel a slight pressure or weight. Remember to stay grounded, centered, and respectful.

The morphic field will undergo a series of still points (CST language), during which you will feel a pulling, twisting, and then settling of the field. Just wait. It may start unwinding again. Be slow and patient. With a problematic situation, it may be a long wait. The final balancing will feel like a rhythmic back-and-forth movement. Let that be felt for a few minutes to ensure it doesn't need to unwind further.

The friend or client will continue to tell you their sensations. They may experience heat flashes, sweaty hands, recall a memory, or have any other unexpected reaction. You may also either empathize with the release of heat or feel it as it leaves their body. If they are abreacting, you can either gently return the field to the body or begin working with them (see the trauma chapter). If you are practiced, you can do both. If not, return the morphic field to the body, letting it know that you can work with it again later. Mentally think this as you gently waft the field back into the body.

If no abreaction has occurred, you will sit with the field in your hands until you feel a prolonged period of back-and-forth balance, as if the field is breathing, allowing for a relaxed and comfortable experience. This is when the field has released its energetic blockage(s).
Sit with your client or friend until they are settled. Hold the room in Sacred Space while all gets back to peace. Of course, all of this has been done with you in your Heart Space. YOU CANNOT DO THIS WITH YOUR BRAIN! You will never be able to think of this procedure as anything but a necessary evil. Trust your hands and your heart to the work, and you will never be disappointed.

If you even begin to try to figure it out, you will leave your heart space and lose the connection. Practicing meditation daily will help you stay centered. (Yoga, massage work done quietly, or other spiritual practices are also good practices to maintain your heart space easily)

When working with a client, I usually have an obvious stare in my eyes. I have dual attention. I am fully aware of my client and, in addition, lock my eyes onto something in the room so that I can sense the energies I am working with. I usually see my client with my peripheral vision. If they need grounding, I bring my eyes to theirs to help them reconnect with their body. I may have to ask them to examine my eyes, and again, I am still working and have split attention between my client and the work that is happening. It has become easy for me to switch to alpha brain waves and have dual attention. With practice, you will be able to do it as well.

About ten years ago, when I was holding my new grandchild in a state of bliss, I saw in my mind's eye her stomach and colon. It was such a clear picture that I could even see the interstitial fluids moving around those organs. I immediately pushed that vision out of my mind as I didn't want to invade this baby's body. What I found out later is that she had had stomach problems for many years. After she would eat, the cramps would come. When she was old enough and with her parents' permission, I sat her down. I explained the concept of the morphic field in her stomach, and that I would allow it to come into my hands to relax and unwind.

After I did this with her a few times one weekend, she has had no more stomach problems. The next morning, she said, "Gramma, can we take out my stomach again? That was so cool!" At this point, I remembered what had happened when she was a baby. It was her body asking me to work with the stomach then, but not being confident that it was OK for me to do, I didn't do it. It is essential to trust your instincts, as being unsure can lead to more harm than good. It was perfect that we got to clear this up so many years later.

Let's say a client is going to need gallbladder surgery. It is never my intention to tell any client that they don't need medical care after I do any energy work. When you perform energy work before the surgery, it allows the energy field of the GB or the affected area or organ to be energetically balanced so that the surgery will be unencumbered by those blockages. You can also schedule a session sometime after the recovery period from surgery to address any issues that may have arisen during the procedure.

Phone Morphic Work

The same standards apply; the client must sense the changes in their energy fields and report them to you while you work. I set this up with a hands-free phone while sitting at my massage table in my treatment room. My arms are resting on the table so as not to strain myself if I need to hold a field for a long time. I mentally perform a body scan or have my client tell me what they would like to work on. I also tell them how I am positioning their body and where my hands will be whenever I make a change in position.

Skype work, where you can see the client, is also helpful since you can more physically direct your hands to specific areas, and the client can see you as well. You are also able to see their facial and bodily expressions clearly.

Example

In phone work, it is crucial that you can mentally visualize what you are working on or at least imagine it. As always, the client must keep you apprised of their sensory experiences. If they can be grounded and relaxed, having a hands-free phone is ideal.
In this example, we will work with a colon. The client may be reporting pain or difficulty with elimination.

Please refer to your anatomy book to familiarize yourself with the structures you will be examining. You may be drawn to the small intestine, the iliocecal valve, the large intestine, the splenic or hepatic flexures, the jejunum, duodenum, bile duct, and so on. The client may report 'stomach', so you must ask if the area is high up under the rib cage (stomach) or lower under the navel (colon). Is it on the left or right side and ask about the type and level of pain or problem. For me, I scan and find. At some point, you will be able to do this also, especially if you study Jose Silva's body scanning procedure.

Sometimes, where the pain IS is not where the problem is. It could be as simple as a stuck ileocecal valve. Now you understand why you need some knowledge of anatomy.

Let us make this case simple. The colon is blocked in one area because the client is not drinking enough water or eating foods that clog the colon. You will find that sometimes you receive pertinent information like this from the organ itself; it just appears in your mind. I am on the phone with the client. As I scan slowly with my mind and/or my hands with an energetic duplicate (morphic field) of their colon area in front of me or in my mind's eye, I am stopped at the point where I need to work.

You will find, after a while, that you literally "run into" the energetic blockage as you scan. Your hand may even feel so pulled to an area that it will feel 'off' if you even begin to move away. You may decide to do several things. You can imagine putting your eight fingers into the client's abdomen underneath the blockage and wait for changes, or you can remove the morphic field and hold the part of the blocked colon in your hand. What I usually do with colon blockages is use the energy rays coming from my fingers to contact and begin to move and break up the blockage. I scan and ensure that the area after the blockage is clear enough to allow for the movement of the blockage.

You will learn precisely how to move your fingers (with those long, extended energy fields entering the area) to begin loosening the affected area. You can even mentally run water through the area to help move it. This water, of course, is only in your mind, but working with etheric fields on etheric fields is quite effective. Remember, thoughts are things. You sending water to the area will truly bring about a change. From here on, it is up to you to determine how long you need to work or what else is required. There is a real sense of **completion** when, in fact, you are done and finished with the work. Being in your Heart Space will allow you to know this inherently with each session.

There certainly can be emotional reasons for this blockage. Ensure your hearing, **sight** (both mentally and physically), and other senses are alert for any signs of breath intake, weeping, or a heavy feeling, as well as an abrupt silence coming through the phone while you are working.

Craniosacral Therapy

You mustn't work in the head of a client unless you know extended cranial anatomy. There are 12 cranial nerves in the brain area, not to mention the medulla oblongata area or brain stem, which is the place we live and breathe from. If you are unfamiliar with the area, please refrain from working there. I spent a few minutes on a client's head by phone, but she couldn't feel the work, so I stopped. That night, she was very nauseated. I had disturbed the vagus nerve, which runs through the gut and diaphragm, controlling digestion, the heart, and other functions. I did not correct it at the time because she could not feel that something was off until she became sick to her stomach. You can see how important it is to get feedback when working energetically.

Therefore, this book will not cover working in the head. Great care is advised in all areas where nerves are directly addressed. If you experience headaches, avoid activities that involve working around the head. Still, please refrain from venturing into the deep recesses of nerves and brain tissue, even morphologically.

Morphic Field Organ Work

As I discovered with my client Karen, you can continue to work with the morphic field of an organ even after it has been surgically removed. The morphic field will remain with all its blocked energies.
Karen was a regular massage client after having had her fourth car accident. At that time, I knew she was holding trauma, as many accidents like she had had, confirmed trauma in the body.

At one session, I told her I would proceed with just energy work. I had her lie on my massage table, clothed, and I stood outside the room and dropped down into my heart. I walked to the doorway and slowly proceeded into the room. About three feet from her body, I felt a firm boundary that said to me, "Go no further". I held my hands toward her body and was pulled to the liver gallbladder area and began feeling and working with the energies.

Within a few minutes, I happened to glance at her face, and she was pale and breathing shallowly. At the same time, I had noticed a strong picture in my mind of green sludge pouring out of her gallbladder area and hitting the floor. It was dramatic. In an instant, I realized that she had lost too much etheric matter, and it was putting her in shock. I closed the area and called her name to open her eyes.

As she did, I said I was going to stand in the corner of the room, away from her, and that she was going to settle down and feel better soon. (Suggestions can be constructive in any session. This comes from understanding hypnosis. In that moment, I saw a huge angel standing at her head, and I asked her if she liked angels. "Oh, yes! I always bring in my angels and use them for protection." I told her that she had a giant one standing next to her, helping her to settle.

Within a few minutes, she was back to normal. Her color had returned, and she was able to sit up easily. I knew she was holding onto anger and bitterness over whatever had happened to her. Even though she was only in her early 20s, she had had her gallbladder taken out because of all the stones. She was angry to the point of creating a physical problem in her body.

The next time I saw her, she shared some instances of recently encountering certain people in her life who had caused her pain, and she was able to be present for that and express how they had impacted her. She mentioned she felt as if she had new breath in her body and was able to move forward.

Yes, Karen needed more help. That was up to her, as I was working with her in one capacity, doing the repair work from her car accident. Suppose a therapist becomes everything for a client, doing too many avenues of healing. In that case, it can also become abusive as a power play on the therapist's part. I do hope she found someone to help her through the rest of her healing work.

You have read some session work in this book that probably sounds amazing. Please note that it is not typically just one session that brings resolution. It is the client who does the work for many days and even months after the work begins. After the light goes on in the mind of the client, sometimes they have to climb out of a bottomless hole they have dug to get away from their trauma. There is no getting away; there is only facing and embodying the trauma, so it no longer has power over you. Even though we cannot 'see' the morphic fields of trauma that surround and run through our bodies, they can be powerfully crippling both physically and emotionally to the point of limiting our very life force.

Remember that if the morphic field is in a state of anxiety or dis-ease, and is held long enough in such a distorted pattern, it can create difficulties in the physical tissues as well. The morphic field holds the pattern; the physical tissues express the pattern.

Chapter 9

Outlining the Work of Trauma Resolution Instructions

I will say again that Babette Rothchild's book, **The Body Remembers,** is excellent, clearly explaining how she works with trauma resolution in sessions. Here, I will begin to share some experiences and practices that may be beneficial for you in your initial practice of trauma resolution.

Please read most of the books I have recommended, especially those by Peter Levine and Ms. Rothchild. And, most importantly, have your own personal work done, or at least be working with a therapist who can also serve as a consultant when you encounter a difficult situation with a client. Even if you do not use formal hypnosis in your sessions, please know that hypnosis itself is always going on. It is essential to understand something about it to utilize it as yet another tool in your work.

To begin, create a sacred space, whether it is in your office or home, where it will be quiet and uninterrupted. On occasion, there have been interruptions, and I count them as just Divine Interference that must serve some purpose. Still, my ability to know is not always apparent, making it OK. Of course, in hypnosis, you can suggest that noises will only allow the subject to go deeper into the state of hypnosis.

You have now settled your client in your treatment room. I like to let them know the procedure or general progression we will be going through to ensure a feeling of safety. I explain that they may feel sensations in their body as they tell their stories. Then we have a demonstration of what a sensation is, not a feeling (such as anger or sadness), but a sensation, such as sweating, heat, or changes in breathing patterns.

I will also observe and name what I see during the session, as well as ask them about their sensations. I ask them to gently scratch their arm and explain that it is a sensation, and to please report it verbally when they experience it during the session.

I let them know that I will interrupt them during the story they are telling when I see that they might be reacting. I also ask them to tell the story slowly. When we stop at a point of activation, I ask them to report sensation. We patiently wait together while it integrates and becomes less. Remember, in my trauma chapter, I explained how to keep the client grounded by maintaining eye-to-eye connection or gently touching their feet (with prior permission) to help them regain their grounding.

If they need a break, we will stop and discuss something else until I see that they are balanced again. When we revisit the part that caused them so much sensation, we both notice that the sensations have lessened. That is because the body had done the integration while we were chatting about an innocuous subject as a short segue before moving on again.

A word of caution here: if you are not a classically trained therapist or have not been trained in Biodynamics or Somatic Experiencing, please refrain from taking on heavy cases.

There is a degree of trauma at which I draw the line and refer out to a more trained therapist. PTSD from war trauma, obviously being one situation, kidnapping, being held hostage, ongoing trauma for years, gang rape, and other rapes, depending on the situation the client is now experiencing. Remember the story of the young woman who was raped in a foreign country by a man holding a machine gun? I referred her to a psychotherapist in my area. It is essential, too, to have a list of professionals for such references.

If you find yourself working with multiple clients per day or week, you should see a therapist you can work with. There is always a small amount of trauma that can be felt and held by the therapist in any session, and it is essential to be aware of this and take care of yourself.

I remember in one of Rava Shavram's instructional CDs (from SpiritRoc.com), he was working with a Grief Counselor who had been activated by a client's issue with the death of one of his parents. Since she had problems with her parents, it brought up all her stuff.

You must recognize when this happens (remember transference and countertransference) so that you do not exacerbate your client's problems or stress yourself to the point of being unable to work effectively with this or any other client. Transference and countertransference can be a valuable tool for therapeutic discovery when attended to.

You can take one incident and work with it, resolving it, and not needing to see that client again. That is more of what I do. On occasion, I will work with a client for an extended period. Still, in between times, they are reading the books I recommend and doing their own work, whether it's having to deal with changing their victim status, or setting boundaries and beginning to understand how powerful being in choice is. I encourage my clients to be empowered and let them know repeatedly that I am not doing anything but opening doors for them; it is their job to go through those doors and make their own changes.

We can have brief phone calls and/or emails (with clear boundaries) to share their successes or occasionally answer questions, but not regular, ongoing weekly sessions.

If I am requested to see someone weekly for an extended period of time, I know I am out of my scope of practice and encourage a referral to a trained therapist. However, if my client has successfully overcome one hurdle after a few months and wants to tackle another, I will see them again. It is all a matter of judgment on your part. Learn to refer out and say no when necessary.

****

SOME BOOKS & FILMS YOU MAY FIND
INTERESTING

Abbot, Edwin: **Flatland, A Romance of Many Dimensions**

Aron, Elaine: **The Highly Sensitive Person**

Barral, Jean-Pierre, **Understanding the Messages of Your Body; Manual Thermal Evaluation**

Basiago, Andrew D. See www.projectpegasus.net as well as YouTube searches on his name

Becker & Selden: **The Body Electric**

Behary, Wendy T., LCSW: **Disarming the Narcissist Surviving & Thriving with the Self-Absorbed**

Bradshaw, Dr. John: see all of his books, CDs on www.johnbradshaw.com

Cantor, Carla and Brian A. Fallon, M.D.: **Phantom Illness: Recognizing, Understanding, and Overcoming Hypochondria**

Carson, Rick: **Taming Your Gremlin** and other books

Chodron, Pema: **The Places that Scare You,** also **Start Where You Are**

Conger, John P.: **The Body in Recovery**

Davidson, Richard J., Ph.D., and Sharon Begley: **The Emotional Life of Your Brain**

Erickson, Milton: (check Amazon for any books about this man; his expertise in working with the subconscious through hypnosis is extraordinary and informative)

Faber, Adele & Mazlish, Elaine: **How to Talk so Kids will Listen and Listen so Kids will Talk**

Forrest, Lynne: **Guiding Principles for Life Beyond Victim Consciousness**

Frankel, Victor: **Man's Search for Meaning**

Goleman, Daniel: **Emotional Intelligence**

Greene, Brian: check Amazon for any of his many books on understandable Quantum Physics, especially **The Hidden Reality: Parallel Universes and the Deep Laws of the Universe**

The HeartMath Institute www.heartmath.org

Herman, Judith: **Trauma and Recovery**

Hass, Rue Anne, M.A.: **The 8 Master Keys to Healing What Hurts for the highly sensitive person** (check

Amazon or her website for her set of 4 books on EFT for the Highly Sensitive Person)

Kaku, Michio: **Parallel Worlds: A Journey Thru Creation, Higher Dimensions, and the Future of the Cosmos**

Kalsched, Donald: **The Inner World of Trauma**
Levine, Peter: Waking the Tiger**, as well as his book and CD on** Healing Trauma: **and** *Trauma Through a Child's Eyes: Awakening the Ordinary Miracle of Healing* **by Peter A. Levine, Ph.D., and Maggie Kline: and** Sexual Healing, Transforming the Sacred Wound CD **Set, and** In an Unspoken Voice: How the Body Releases Trauma and Restores Goodness

Lipton, Bruce: **The Biology of Belief**

Motz, Julie: **Hands of Life**

Oxnam, Robert B.: **Fractured Mind: My Life with Multiple Personality Disorder**
Porges, Stephen W.: **The Polyvagal Theory: Neurophysiological Foundations of Emotions, Attachment, Communication, and Self-Regulation**
Roberts, Jane: **The Nature of Personal Reality, A Seth Book**

Rothschild, Babette: **The Body Remembers** (also see www.bodynamics.com for training in this body-centered psychology)

Sarno, John E.: **The Mind Body Prescription: Healing the Body, Healing the Pain**

Siegel, Daniel: **The Developing Mind**

Silva, Jose: **The Silva Mind Control Method; You the Healer**

Sheldrake, Rupert: **A New Science of Life** (about Morphic Fields), also see his other books.

Talbot, Michael: **The Holographic Universe**

Taylor, Eldon www.eldontaylor.com

Vander Kolk, McFarlane, Weisaeth: **Traumatic Stress** Villoldo, Alberto, Ph.D. **Mending the Past and Healing the Future with Soul Retrieval**

Von Franz, Louise: **Creation Myths**
What the Bleep: Through the Wormhole
(Movie on DVD)

Wolf, Fred Alan: **The Dreaming Universe** (and others) (He is also featured in the DVD, **What the Bleep**

TED Talks on YouTube is one of the best places for short inspirational talks. Currently, I listened to Dr. Brene Brown on Shame and Vulnerability. It was excellent!! There are many more to check out as well.